CONCEPT AND APPLICATION OF
SHARIAH FOR THE
CONSTRUCTION INDUSTRY

Shariah Compliance in Construction Contracts,
Project Finance and Risk Management

CONCEPT AND APPLICATION OF
SHARIAH FOR THE
CONSTRUCTION INDUSTRY

Shariah Compliance in Construction Contracts,
Project Finance and Risk Management

Editors

Khairuddin Abdul Rashid
International Islamic University Malaysia, Malaysia

Kiyoshi Kobayashi
Kyoto University, Japan

Sharina Farihah Hasan
International Islamic University Malaysia, Malaysia

Masamitsu Onishi
Kyoto University, Japan

World Scientific

NEW JERSEY · LONDON · SINGAPORE · BEIJING · SHANGHAI · HONG KONG · TAIPEI · CHENNAI · TOKYO

Published by

World Scientific Publishing Co. Pte. Ltd.

5 Toh Tuck Link, Singapore 596224

USA office: 27 Warren Street, Suite 401-402, Hackensack, NJ 07601

UK office: 57 Shelton Street, Covent Garden, London WC2H 9HE

Library of Congress Cataloging-in-Publication Data
Names: Khairuddin, Abdul Rashid, editor.
Title: Concept and application of Shariah for the construction industry : Shariah compliance in
    construction contracts, project finance and risk management / Editors: Khairuddin Abdul Rashid,
    International Islamic University Malaysia, Kiyoshi Kobayashi, Kyoto University, Sharina Farihah
    Hasan, International Islamic University Malaysia, Masamitsu Onishi, Kyoto University.
Description: Hackensack, N.J. : World Scientific, 2018. | Includes index.
Identifiers: LCCN 2018015080 | ISBN 9789813238909 (hardcover)
Subjects: LCSH: Construction contracts (Islamic law) | Dispute resolution (Law)
Classification: LCC KBP893.3.C65 C66 2018 | DDC 624.068/1--dc23
LC record available at https://lccn.loc.gov/2018015080

British Library Cataloguing-in-Publication Data
A catalogue record for this book is available from the British Library.

For any available supplementary material, please visit
https://www.worldscientific.com/worldscibooks/10.1142/10957#t=suppl

Desk Editor: Tay Yu Shan

Typeset by Stallion Press
Email: enquiries@stallionpress.com

Printed in Singapore

Contents

Acknowledgments   ix

About the Editors   xi

Contributors' Biographies   xiii

Preface   xix

List of Tables   xxvii

List of Figures   xxix

List of Text Boxes   xxxi

List of Appendices   xxxiii

List of Abbreviations   xxxv

List of Surahs (Chapters) in the Quran   xxxix

Part 1:   Shariah   **1**

1   Shariah and Its Meaning in Islam   3
      M. KAMAL Hassan

2   Understanding of the Shariah in Regards to Construction   15
     MOHAMAD AKRAM Laldin

3   Promoting Efficiency in Construction Practices:
     Lessons from Shariah   39
     AINUL JARIA Maidin

4   Shariah-Compliant Construction Marketing:
     Development of a New Theory   71
     KHAIRUDDIN Abdul Rashid and Christopher
     Nigel PREECE

Part 2:   Shariah-Compliant Construction Contract:
            Concept and Application   **81**

5   Shariah-Compliant Contract: Concept and
     Application for Construction Works   83
     KHAIRUDDIN Abdul Rashid

6   *Istisna'* Model for Construction Works' Contracts   107
     KHAIRUDDIN Abdul Rashid

7   Validity of *Istisna'* for Construction Works Contracts   121
     KHAIRUDDIN Abdul Rashid

8   The Application of Limited Liability in Construction
     Contracts from the Malaysian Law and Shariah
     Perspectives   129
     ZUHAIRAH ARIFF Abdul Ghadas

9   The Application of Shariah Principles of ADR
     in the Malaysian Construction Industry   145
     ZUHAIRAH ARIFF Abdul Ghadas, ROZINA
     Mohd Zafian and ABDUL MAJID TAHIR Mohamed

**Part 3:   Shariah-Compliant Project Finance
and Risk Management   167**

10   The 3 Rs in Islamic Project Finance: Its Relevance
Under *Maqasid al-Shariah*   169

Etsuaki YOSHIDA

11   Islamic Home Financing Through *Musharakah
Mutanaqisah*: A Crowdfunding Model   181

AHAMED KAMEEL Mydin Meera

12   Proposed Model on the Provision of Affordable
Housing via Collaboration Between
Wakaf-Zakat-Private Developer   205

*KHAIRUDDIN Abdul Rashid,
SHARINA FARIHAH Hasan and
AZILA Ahmad Sarkawi*

13   Shariah Compliance Risk-Sharing in Islamic Contracts   223

Masamitsu ONISHI and Kiyoshi KOBAYASHI

14   *Takaful* for Construction Works' Contract:
Concept and Application   253

*PUTERI NUR FARAH NAADIA Mohd Fauzi
and KHAIRUDDIN Abdul Rashid*

15   *Hibah Mu'Allaqah* (Conditional Gift) and
Its Application in *Takaful*   271

AZMAN Mohd Noor

16   Managing the Risk of Insolvency in the Construction
Industry with Equity Financing: Lessons from *Nakheel*   287

ABDUL KARIM Abdullah

Index   299

Acknowledgments

We would like to express our gratitude to Allah ($s.w.t$) for it is only with His pleasure and blessings that we are able to complete the tasks of selecting, compiling, editing and publishing this book.

We are indebted to many individuals and institutions for assisting us in turning this book into a published reality. It would not be possible to mention all but their assistance is very much appreciated.

In particular we would like to thank the Rector and management of the International Islamic University Malaysia (IIUM) and the President and management of Kyoto University of Japan for their approval to organize a series of IIUM-Kyoto University Research Colloquiums on Shariah-Compliant Issues in Construction during the period between 2013 and 2015. Talented and quality papers presented during the event were developed and have become the chapters of this book.

Our thanks are due to the members of the Organizing Committee of the IIUM-Kyoto University Research Colloquium on Shariah-Compliant Issues in Construction 2013–2015. Their effort in staging the event led to the publication of this book.

Our special thanks are due to the authors for consenting to have their works edited and published.

The following individuals assisted the Editors in preparing the manuscript: Dr. Suazlan Mt Aznam for the initial formatting,

Ms. Zakiah Abdul Karim for the initial proofreading, Dr. Puteri Nur Farah Naadia Mohd Fauzi and Dr. Mahadi Ahmad for the proofreading of the Quranic verses and *hadith*s, Mr. Hasri Hasan and Mr. Chua Hong Koon for their assistance in organizing the publication.

Thank you all.

Editors,

Professor Sr. Dr. KHAIRUDDIN Abdul Rashid
International Islamic University Malaysia

Professor Dr. Kiyoshi KOBAYASHI
Kyoto University Japan

Assistant Professor Dr. SHARINA FARIHAH Hasan
International Islamic University Malaysia

Associate Professor Dr. Masamitsu ONISHI
Kyoto University Japan

January 2019

About the Editors

KHAIRUDDIN Abdul Rashid, Sr. Dip., BSc., MSc., Ph.D., MRISM, CQS (BQSM) is Professor at the Department of Quantity Surveying, Kulliyyah of Architecture and Environmental Design and Head, Procurement and Project Delivery System Research Unit, International Islamic University Malaysia (IIUM). He is co-founder and co-editor of the *Journal of Quantity Surveying and Construction Business* and reviewer of several international refereed journals related to construction. Professor Khairuddin has authored, co-authored, edited and co-edited 14 books and more than 200 technical papers on quantity surveying, procurement, contract administration including Shariah-compliant and Public Private Partnership (PPP) or Private Finance Initiative (PFI).

Kiyoshi KOBAYASHI, BSc., MSc., Eng.D., is Professor at the Graduate School of Management as well as Graduate School of Engineering, Kyoto University. He is a renowned researcher in the fields of Infrastructure Management and Regional Science and a recipient of several awards including the Distinguished Research Awards by Japan Society of Civil Engineers (JSCE) and Fellow Awards by RSA International. He is the President of JSCE, and serves on the editorial boards of international journals including the *Journal of Infrastructure Systems*. He is the author and co-editor of 65 books and over 430 academic reviewed papers.

SHARINA FARIHAH Hasan, Dip., BSc., MSc. (Eng.), Ph.D. (Eng.), is Assistant Professor at the Department of Quantity Surveying, Kulliyyah of Architecture and Environmental Design, IIUM. She is an editorial board member of the *Journal of Quantity Surveying and Construction Business*. Dr. Sharina has authored and co-authored more than 40 papers in books, book chapters, journal articles and proceedings on quantity surveying, prequalification of contractors, labor market and productivity, and Shariah compliance in construction.

Masamitsu ONISHI, BSc., MSc., Eng.D., is Associate Professor at the Disaster Prevention Research Institute, Kyoto University. He is interested in institutional systems related to the procurement of infrastructure services including contracting and procurement schemes particularly PPPs and disaster risk governance. He is a recipient of the JSCE Young Researcher Award 2006 and Best Paper Award 2008. He has authored and co-authored more than 50 refereed papers and 12 books, chapters in books and proceedings.

Contributors' Biographies

ABDUL KARIM Abdullah (Leslie TEREBESSY), B.A. (Hons), M.A., M.Ed. is President of Terebessy Foundation, a think-tank based in Toronto, Canada (www.terebessyfoundation.com). He taught economics for 10 years in Canada and Malaysia. He published works on Islamic finance and economics. He co-edited *Islamic Finance: Issues in Sukuk and Proposals for Reform*, Islamic Foundation, UK, 2014, and *Islamic Studies in World Institutions of Higher Learning*, USIM, Malaysia, 2006. He is author of *Issues in Islamic and Conventional Finance: A Critical Appraisal*, forthcoming.

ABDUL MAJID TAHIR Mohamed, LL.B., Ph.D. is lecturer and Deputy Dean (Academic and Graduate) at the Faculty of Law and International Relations, University Sultan Zainal Abidin, Terengganu, Malaysia.

AHAMED KAMEEL Mydin Meera, B.A., M.A., Ph.D. is Professor and currently Managing Director of Z Consulting Group Sdn. Bhd. He has taught, researched, published and consulted in areas including Islamic Finance, Economics, Money and Payment Systems; and is particularly known for his writings on real money systems. He authored the books *The Islamic Gold Dinar*, Pelanduk Publications, 2002; *The Theft of Nations*, Pelanduk Publications, 2004; and *Real Money*, IIUM Press, 2009.

AINUL JARIA Maidin, LL.B (Hons), LL.M, Ph.D. is Professor at the Ahmad Ibrahim Kulliyyah of Laws, IIUM. Her research interests are legal and administrative framework in the areas relating to sustainable development, disaster risk management, land administration and management, property, building construction and maintenance. She has authored and co-authored 60 books, chapters in books and conference proceedings.

AZILA Ahmad Sarkawi, B.A. (Hons), M.A., Ph.D. is Associate Professor at the Department of Urban and Regional Planning, IIUM. Her areas of specialization are planning and environmental laws and Islamic built environment. She received the Planning Scholar Award twice in 2009 and 2016. She has authored and co-authored more than 90 papers related to her areas of specialization.

AZMAN Mohd Noor, B.A. (Hons), M.A., Ph.D. is Associate Professor at the International Institute of Islamic Banking and Finance, IIUM. His research interests are on Islamic banking products, *takaful* (Islamic insurance) and Islamic capital market. He is Shariah committee member at a number of Islamic financial institutions in Malaysia.

Christopher Nigel PREECE, BSc., Ph.D. is Professor of Project Management at the College of Engineering, Abu Dhabi University. For 15 years Professor Preece was with the Construction Management Group at the School of Civil Engineering, University of Leeds, UK. From 2008 to 2010, he was with the Department of Quantity Surveying at the IIUM and from 2010 to 2015, he was with the Razak School of Engineering and Advanced Technology, Universiti Teknologi Malaysia. He is a recognized authority in international construction, project and business development and marketing. He is a Fellow of the Chartered Institute of Building and an international Construction Ambassador for the Chartered Institute of Building (CIOB). He is a Chartered Marketer and a Fellow of the Higher Education Academy, UK.

Etsuaki YOSHIDA, Eng.D. is Project Associate Professor at the Graduate School of Asian and African Area Studies of Kyoto University. He is also Director and Senior Economist at Japan Bank for

International Cooperation. His research interests are international finance, emerging economies, and Islamic finance from a practical point of view, including FinTech, macro-financial stability and product development. He has authored and co-authored five books on Islamic finance in Japanese and book chapters. Dr. Yoshida is a frequent speaker on Islamic finance in events including World Islamic Banking Conference, Islamic Financial Services Board (IFSB) seminar, and the *Sukuk* summit.

KHAIRUDDIN Abdul Rashid, Sr. Dip., BSc., MSc., Ph.D., MRISM, CQS (BQSM) is Professor at the Department of Quantity Surveying, Kulliyyah of Architecture and Environmental Design and Head, Procurement and Project Delivery System Research Unit, IIUM. He is co-founder and co-editor of the *Journal of Quantity Surveying and Construction Business* and reviewer of several international refereed journals related to construction. Professor Khairuddin has authored, co-authored, edited and co-edited 14 books and more than 200 technical papers on quantity surveying, procurement, contract administration including Shariah-compliant and Public Private Partnership (PPP) or Private Finance Initiative (PFI).

Kiyoshi KOBAYASHI, BSc., MSc., Eng.D. is Professor at the Graduate School of Management as well as Graduate School of Engineering, Kyoto University. He is a renowned researcher in the fields of Infrastructure Management and Regional Science and a recipient of several awards including the Distinguished Research Awards by JSCE and Fellow Awards by RSA International. He is the President of JSCE, and serves on the editorial boards of international journals including the *Journal of Infrastructure Systems*. He is the author and co-editor of 65 books and over 430 academic reviewed papers.

M. KAMAL Hassan, B.A., M.A., MPhil., Ph.D. is Distinguished Professor, International Islamic University Malaysia (IIUM). He joined IIUM since its inception in 1983, becoming the Founding Dean of the Kulliyyah of Islamic Revealed Knowledge and Human Sciences, Deputy Rector for Academic Affairs and Rector. He specializes in contemporary Islamic thought with special reference to Southeast Asia. His publications focus on the concept of integrated education, islamization of human knowledge, Islamic ethics in contemporary society and the concept of Islamic moderation in the Malay world.

Masamitsu ONISHI, BSc., MSc., Eng.D. is Associate Professor at the Disaster Prevention Research Institute, Kyoto University. He is interested in institutional systems related to the procurement of infrastructure services including contracting and procurement schemes particularly PPPs and disaster risk governance. He is a recipient of the JSCE Young Researcher Award 2006 and Best Paper Award 2008. He has authored and co-authored more than 50 refereed papers and 12 books, chapters in books and proceedings.

MOHAMAD AKRAM Laldin, B.A. (Hons), Ph.D. is Professor and currently the Executive Director of International Shari'ah Research Academy for Islamic Finance (ISRA), Malaysia. His research interests are Islamic banking and finance, Islamic capital market, *takaful* and fundamentals of Islamic jurisprudence. He was recipient of the Zaki Badawi Award for Excellence in Shariah Advisory and Research in 2010 and recipient for Most Outstanding Individual Contribution to Islamic Finance during the Kuala Lumpur Islamic Finance Forum (KLIFF) in 2016. He has authored and co-authored several books.

PUTERI NUR FARAH NAADIA Mohd Fauzi, BSc., Ph.D. is Assistant Professor at the Department of Quantity Surveying, Kulliyyah of Architecture and Environmental Design, IIUM. She is a Graduate Member of Board of Quantity Surveyors Malaysia (BQSM) and Registered Member of Royal Institution of Surveyors Malaysia (RISM) and a member of the Procurement and Project Delivery System Research Unit at IIUM. She has published works on *takaful* for construction works, Shariah-compliant construction contracts, procurement and quantity surveying.

ROZINA Zafian, BSc., MSc., Ph.D., MRISM, BQSM is Senior Superintending Quantity Surveyor at the Contract and Quantity Surveying Division, Public Works Department, Malaysia.

SHARINA FARIHAH Hasan, Dip., BSc., MSc. (Eng.), Ph.D. (Eng.) is Assistant Professor at the Department of Quantity Surveying,

Kulliyyah of Architecture and Environmental Design, IIUM. She is an editorial board member of the *Journal of Quantity Surveying and Construction Business*. Dr. Sharina has authored and co-authored more than 40 papers in books, book chapters, journal articles and proceedings on quantity surveying, prequalification of contractors, labor market and productivity, and Shariah compliance in construction.

ZUHAIRAH ARIFF Abd Ghadas, LL.B (Hons), LL.M, Ph.D. is Professor at the Faculty of Law and International Relations, University Sultan Zainal Abidin, Terengganu, Malaysia. Her area of specialization is in Business Law, particularly on development of business entities. She has won gold medals in innovation and research competitions at the university level and silver medals at the International Invention, Innovation and Technology Exhibition (ITEX) in 2015 and 2017.

Preface

About the Book

Concept and Application of Shariah for the Construction Industry: Shariah Compliance in Construction Contracts, Project Finance and Risk Management is an edited work, carefully chosen from the proceedings of a series of international colloquium on Shariah compliance in construction works. The colloquiums, held between 2013 and 2015, were jointly organized by the International Islamic University Malaysia and Kyoto University, Japan.

The book, consisting of 16 chapters, is structured in three parts:

- Part 1 — Shariah. Presented in four chapters, Part 1 discusses the basic meaning of Shariah and its application for business transactions including in construction. Being definitive in its orientation, Part 1 serves as foundation for the ensuing discussions in Parts 2 and 3.
- Part 2 — Shariah-Compliant Construction Contract: Concept and Application. Presented in five chapters, Part 2 is the book's mainstay. It discusses the Shariah and its application for construction, focusing on contracts for construction works. This part of the book also contains proposals and innovative models on how Shariah compliance could be applied for construction works contracts.
- Part 3 — Shariah-Compliant Project Finance and Risk Management. Presented in seven chapters, Part 3 discusses the Shariah

approach in project finance and risk management and proposes ideas on their application for construction.

The Editors aspire for the book to achieve the following objectives:

- To share with students, researchers, practitioners and policymakers of construction and its related businesses and activities on basic theories, contemporary and innovative ideas on Shariah compliance and its application for construction. In this regard, the interest of the Editors is to facilitate readers in enhancing their understanding; and subsequently
- To form the basis for further and in-depth research, publication and application of Shariah compliance in construction.

Context

Islam calls for transactions between people — mutual dealings including in businesses — to be performed in accordance with the Shariah. Within the Shariah the broad principles that are applicable is *al-muamalat*. Al-*muamalat* (singular: *muamalah*) emphasizes the need for business transactions to apply the concepts of justice, moral obligation, accountability and equality; these aspects are in line with the Islamic belief (*iman*), practices (*amal*) and value system.

The last two decades saw the term "Shariah-compliant" and its application in business transactions increasing in popularity. The Asian financial crisis of 1997, global financial crisis of 2007–2008, the Enron scandal and other reported ills besieging the Western-style business transactions led to advocates of Shariah-compliant business transactions promoting it as a credible alternative.

Consequently, Islamic and other scholars, experts, researchers and students conduct research and debate on Shariah-compliant business transactions at discourses while practitioners apply the concept and continuously seek ways to improve its delivery. But it seems these activities are more prominent in commerce, banking and finance as opposed to construction. The sluggish uptake by the construction industry could be due to reasons including limited understanding of Shariah itself and the concept of Shariah-compliant, the lack of research and publication on Shariah and its application for construction. This book endeavors to address some of these deficiencies.

Shariah

Islam is *ad-deen* or way of life. The key guiding principles of Islam is the Shariah. It contains guidance pertaining to relationships between man and Allah (*s.w.t*), man and man, and man and the environment.

Shariah is derived from revealed sources i.e. Quran and *as-Sunnah* and non-revealed sources i.e. *ijma'* and *ijtihad*.

- Quran — The book that contains the speech of Allah (*s.w.t*) revealed to the Prophet Muhammad (pbuh) and transmitted to mankind by continuous testimony.
- *as-Sunnah* — Narrations (*hadith*) from the Prophet (his acts, sayings, whatever had been approved) and reports which described his physical attributes and character.
- *Ijma'* — Agreement of opinion among the Muslim scholars after the life of Prophet (pbuh) on problems which are not determined definitely and directly in Quran and *as-Sunnah*.
- *Ijtihad* — Continuous process of reasoning to interpret and harmonize the Divine messages with the changing nature of the life of the Muslim community. The manifestations of *ijtihad* include analogical deductions (*qiyas*), juristic preference (*istihsan*), consideration of public interest (*masalih al-mursalah*), presumption of continuity (*al-istishab*), etc.

The aim of Shariah is to safeguard people's interests in this world and hereafter. It comprises of three key components: faith (*aqidah*), practice or deeds (*amal*) and behavior (*akhlaq*) as illustrated in Figure P.1.

The objective of Shariah or *Maqasid al-Shariah* is the protection of five basic aspects of human life (*al-Kulliyat al-Khams*) i.e. religion, life, mind, property and progeny. Islam views these aspects as the foundation of a harmonious and peaceful society and therefore, transgression against any of these is prohibited.

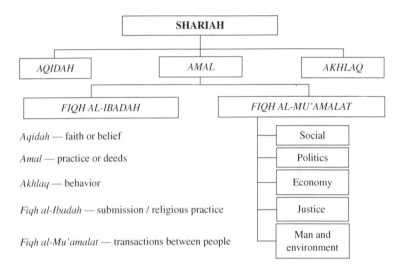

Figure P.1.   Components of Shariah.

There are two kinds of rulings in the Shariah i.e. rulings that are deduced from the decisive evidence in the Quran and *as-Sunnah*, and rulings that are deduced from the *ijtihad* of Muslim scholars. The former are consistent notwithstanding time, place and circumstances, while the latter are flexible and can be varied due to the prevailing time, place and circumstances.

The rulings categorize activities into *fard* or *wajib*, *sunnah*, *haram*, *makruh* and *mubah*. Muslims must adhere to commands that are *fard* or *wajib* and avoid transactions or activities stipulated as *haram*. Table P.1 shows the rulings under the Shariah.

In explaining permissible (*halal*) and non-permissible (*haram*) in Islam, Yusuf Al-Qaradhawi (2013: 17) states:

> "…in Islam, things are prohibited only because they are impure or harmful. If something is entirely harmful, it is *haram*, and if it is entirely beneficial it is *halal*; if the harm of it outweighs its benefit it is *haram*, while if its benefit outweighs its harm it is *halal*."

Shariah-Compliant

Shariah-compliant refers to transactions or activities which are performed according to the rules and standards of the Shariah. In

Table P.1. Rulings Under Shariah.

Sanction (*hukm*)	Definition
Fard or *wajib*	Absolute command which is compulsory and obligatory to be complied with.
Sunnah	Recommended acts to be undertaken by Muslims.
Haram	Action that is prohibited and unlawful.
Makruh	Action of which is preferred to be omitted.
Mubah	Action of which its commission or omission neither merits any rewards nor entails any punishment.

addition, a Shariah-compliant transaction or activity needs to be in line with the objectives of the Shariah (*Maqasid al-Shariah*).

Muslim jurists have identified the key elements that must be avoided in order for a business transaction to be Shariah-compliant (Razali Nawawi, 1999; Muhammad Rawwas, 2005):

- *Iktinaz* (hoarding, black marketing) and *talaqqi al-rukban* (middle person) that lead to deception and inflation.
- *Ihtikar* (monopoly) where a single individual or company is the only supplier of commodities, food or services.
- Transactions involving prohibited subject matters such as trading in alcohol, pork, dead animals and prostitution.
- *Riba* (usury, interest charges) in loans in which a premium must be paid by the borrower to the lender together with the principal amount.
- *Gharar* (uncertainty) when a person undertakes a venture without sufficient knowledge and excessive risk such as gambling.

In addition, a business transaction is considered Shariah-compliant if it falls under the corresponding Islamic legal maxims.[1] There exist an established body of knowledge on Islamic legal

---

[1]Mohamad Akram Laldin *et al.* (2013, pp. ii–iii) highlights the five major Islamic legal maxims, the first relates to "prerequisite to critical thoughts and four that focus on the nature of contracts." They are: (i) Judgment is to be based on knowledge and understanding; (ii) The presumption of validity and permissibility applies to all contracts and conditions; (iii) The fundamental requirement in every contract is justice; (iv) A general principle in contracts is the consent of both parties,

maxims. However, due to the large number of legal maxims and the need for expert guidance in their understanding a discussion on the subject is not within the scope of this book.[2]

One other important aspect of Shariah-compliant is the concept of *halalan toyyiban*. It refers to Shariah-compliant supply chain management wherein a product, say a food item, is designated Shariah-compliant only if the entire processes involved in its production and distribution — from the original sources of the raw materials to point of final consumption — follow the requirements of the Shariah. Thus, it would be inadequate if a business transaction complies with the Shariah but the upstream and downstream activities related to the transaction failed to adhere to the requirements of the Shariah. Consequently, Muslims endeavor to perform Shariah-compliant business transactions that are *halalan toyyiban* in nature.

Shariah Compliance and Its Application for the Construction Industry

Shariah-compliant construction contract is defined as "...contract with its subject matter, agreement, terms and conditions that embrace the Islamic belief, practice and value system." (Khairuddin, 2007).

In the context of construction it is observed that clients and construction industry players have occasionally employed, among others, *bai al-Istisna'* (manufacturing contract) and *sukuk* (equivalent to bond) in financing projects and *takaful* (Shariah-compliant insurance) in insuring works' contracts. Thus, it may be concluded that Shariah-compliant has been applied in construction, albeit in a minimal manner. This conclusion is not to be unexpected as the tremendous performance of Shariah-compliant business transactions have impacted almost all economic sectors including construction.

---

and the effective terms and conditions are what they agree to in contracts; and (v) Deferment constitutes a part of the price.

[2]Readers are advised to consult experts on Islamic jurisprudence or refer to credible publications such as the works of Mohammad Akram Laldin *et al.* (2013), *Islamic Legal Maxims & Their Application in Islamic Finance* and Abdurrahman Raden Aji Haqqi (1999), *The Philosophy of Islamic Law of Transactions.*

However, the application of Shariah compliance and achieving *halalan toyyiban* in construction still have a long way to go. Reasons for these include limited understanding among the key players and researchers of construction about Shariah itself and the concept of Shariah-compliant, and the lack of research and publication on the Shariah and its application for construction.

Given that research and publication on Shariah compliance in construction are lacking, at this point in time, not much can be said about the subject matter. But the deficiency gives rise to opportunity for further and in-depth research to be done. This is especially true as the scope of Shariah compliance in the construction supply chain, from initiation of a project, construction to completion, maintenance, repair, refurbishment, demolition and rebuilding, is very wide. The publication of this book represents a small step towards enriching the Shariah-compliant construction supply chain body of knowledge.

The Editors hope this book acts as a catalyst that would spur deeper interest leading to further research, discourses and publication on the subject of Shariah-compliant and its application for construction.

References

Abdurrahman, Raden Aji Haqqi (1999). *The Philosophy of Islamic Law of Transactions.* Kuala Lumpur: Univision Press.

Khairuddin, Abdul Rashid (2007). Shariah compliant contract: A new paradigm in multi-national joint venture for construction works. In *International Joint Ventures: Reaching Strategic Goals.* K Kobayashi *et al.* (eds.), pp.14–25. Bangkok: Asian Institute of Technology Thailand and Kyoto University, Japan.

Mohamad, Akram Laldin, Said Bouheraoua Riaz Ansary, Mohamed Fairooz Abdul Khir Mohammad Mahbubi Ali and Madaa Munjid Mustafa (2013). *Islamic Legal Maxims & Their Application in Islamic Finance.* Kuala Lumpur: ISRA.

Muhammad, Rawwas Qal'ahji (2005). *Urusan Kewangan Semasa Menurut Perspektif Syariah Islam [Financial Issues from Shariah Perspective].* (Translated by Basri Ibrahim al-Hasani al-Azhari). Kuala Lumpur: al-Hidayah Publishers.

Razali, Nawawi (1999). *Islamic Law in Commercial Transactions.* Kuala Lumpur: CT Publications.

Yusuf, Al-Qaradhawi (2013). *The Lawful and The Prohibited in Islam.* (Translated by Kamal El-Helbawy, M. Moinuddin Siddiqui & Syed Shukry and Reviewed by Ahmad Zaki Hammad). Kuala Lumpur: Islamic Book Trust.

List of Tables

Table P.1   Rulings Under Shariah xxiii

Table 2.1   *Riba* Categories and Rules 23

Table 4.1   Developing the Elements of Shariah-Compliant
             Construction Marketing 77

Table 5.1   Shariah-Compliant Contracts for Business
             Transactions 86
Table 5.2   Comparison Between Shariah-Compliant
             Contract and Conventional-Styled Contracts . . . 91
Table 5.3   Comparison Between Conventional Contract
             Practice and the Shariah (*Istisna'*) 97

Table 11.1   Total Global Debt in Selected Countries and
             Economic Groups 185
Table 11.2   Payments Schedule for *Musharakah Mutanaqisah*
             Partnership 191
Table 11.3   Comparison Between Conventional Loan, BBA
             and MM 192

Table 13.1   Poles of Interpretation of *Gharar* 224
Table 13.2   Payoff Transaction of Lost Camel 229
Table 13.3   Regret Payoff 229

Table 13.4   State-Contingent RR Payoff 231
Table 13.5   Modified Payoff of Lost Camel Transaction 238
Table 13.6   Assessment Result of *Gharar* for Conventional
              Transaction Cases 243

Table 14.1   Conventional Risk Mitigation Methods 255
Table 14.2   Comparison of *Takaful* and Conventional
              Insurance 258
Table 14.3   Summary of Comparative Analysis in Terms
              of the Main Provisions Between a Typical CAR
              Takaful Scheme and CAR Insurance Policy 265

List of Figures

Figure P.1   Components of Shariah xxii

Figure 2.1   The Components of Shariah 17

Figure 5.1   *Istisna'* Financing Contract 95
Figure 5.2   Key Provisions of Typical Standard Forms
              of Contract 96

Figure 6.1   Conventional-Traditional Construction
              Contract 110
Figure 6.2   Basic *Istisna'* Contract 112
Figure 6.3   *Istisna'* Contract Model 1 113
Figure 6.4   *Istisna'* Contract Model 2 — The Parallel
              Istisna' 114

Figure 7.1   A Sample Page of a Typical BQ 126

Figure 10.1   Chronological Development of Major Islamic
              Financial Products 172
Figure 10.2   Full-Recourse and Non-Recourse Loans 175

Figure 10.3   Comparison of Financial Products by
               Factors 176

Figure 11.1   The Mismatch Between Monetary Growth
               and Real Economy Growth 184
Figure 11.2   Islamic Banking and Conventional Banking
               Under Fractional Reserve System 188
Figure 11.3   ETHIS Finances the Affordable Home
               Developer 197
Figure 11.4   ETHOS Finances the Home Buyer 197

Figure 12.1   Proposed *Wakaf-Zakat*-Real Estate Developer
               SPV Model for Affordable Housing 215

Figure 13.1   Contract Game Under Uncertainty 234

Figure 14.1   Key Events on the Origin and Evolution
               of *Takaful* 258
Figure 14.2   Key Development of *Takaful* in the Malaysian
               Takaful Industry 262
Figure 14.3   Key Events Relevant to the Insurance of Works
               Under the PWD 203 264

List of Text Boxes

Box 5.1   *Istisna'* According to the Islamic Development
Bank 95

Box 6.1   International Islamic Academy of *Fiqh*'s Decision
on *Istisna'* 111

Box 6.2   Proposed Model of a More Holistic *Istisna'*
for Construction Works Contracts 118

List of Appendices
(Listed according to Chapter)

Chapter 11

Appendix 200

Chapter 13

Appendix 249

List of Abbreviations

a.s	*'alayhi wasallam*(عليه السلام) — Similar with pbuh
AAOIFI	Accounting and Auditing Organization for Islamic Financial Institutions
ADR	Alternative Dispute Resolution
AHL	Ansar Housing Limited
APR	Annual Percentage Rate
ASEAN	Association of South East Asian Nations
BBA	*al-Bai Bithaman Ajil*
BIM	Building Information Modelling
BLR	Base Lending Rate
BQ	Bills of Quantities
CAR	Contractor's All Risk
CDOs	Collateralized Debt Obligations
CESMM	Civil Engineering Standard Method of Measurement
CIArb	Chartered Institute of Arbitrators
CIDB	Construction Industry Development Board
CIPAA	Construction Industry Payment and Adjudication Act
CIS	Commonwealth of Independent States
D & B	Design and Build
DB/T	Design-build or Turnkey
DBB	Design-bid-build
DBFOM	Design-Built-Finance-Operate-Maintain
DLP	Defects Liability Period

EFS	Export Financing Scheme
ETF	Exchange-traded Fund
ETHIS	Ethical and Islamic Crowdfunding
ETHOS	Equity-Type Home Ownership Scheme
FGD	Focus Group Discussion
FMB	Financial Mediation Bureau
FRB	Fractional Reserve Banking
FSMP	Financial Sector Masterplan
GDP	Gross Domestic Product
IAP	Investment Account Platform
ICAAP	Internal Capital Adequacy Assessment Process
ICC	International Chamber of Commerce
ICHC	Islamic Cooperative Housing Corporation
ICT	Information Communication Technology
IDB	Islamic Development Bank
IEM	The Institution of Engineers Malaysia
IFSA	Islamic Financial Services Act
IPF	Islamic Project Finance
IRR	Internal Rate of Return
ISRA	International Shariah Research Academy for Islamic Finance
JCT	Joint Contracts Tribunal
JKR	Jabatan Kerja Raya
KLRCA	Kuala Lumpur Regional Centre for Arbitration
KPIs	Key Performance Indicators
LAD	Liquidated and Ascertained Damages
LIBOR	London Interbank Offered Rate
lit.	literally
Louis XIV	King Louis the 14th
Ltd	Limited
MAIN	Majlis Agama Islam Negeri-negeri (State Islamic Religious Council)
MIArb	Malaysian Institute of Arbitrators
MLJ	The Malayan Law Journal
MM	*Musharakah Mutanakisah*

OIC	Organisation of Islamic Conference
PAM	Persatuan Akitek Malaysia
pbuh	peace be upon him
PCFC	Ports, Customs and Free Zone Corporation
PFI	Private Finance Initiative
pl.	plural
PPP	Public Private Partnership
PPR	*Program Perumahan Rakyat* (People's Housing Scheme)
PWD	Public Works Department
QAIS	Quality Affordable Islamic Sustainable
R.A	*RadiAllahu 'anhu / 'anha / 'anhuma / 'anhum* (رضي الله عنه). May Allah be pleased with him/her/them
Rev. 1/2010	Revised 1/2010
S.O	Superintending Officer
s.w.t	*Subhanahu Wa Ta'ala* (سُبْحَانَهُ وَ تَعَالَى) (meaning "the most glorified, the most high")
SMM	Standard Method of Measurement
SOCSO	Social Security Organization
SPV	Special Purpose Vehicle
STM	Syarikat Takaful Malaysia
UAE	United Arab Emirates
UK	United Kingdom
UNCITRAL	United Nations Commission on International Trade Law
UNIDROIT	International Institute for the Unification of Private Law
USA	United States of America
USD	United States Dollar
WCED	World Commission on Environment and Development

List of *Surahs* (Chapters) in the Quran

1	Al-Fatihah	24	An-Nur	47	Muhammad
2	Al-Baqarah	25	Al-Furqan	48	Al-Fath
3	Al-'Imran	26	Ash-Shu'ara'	49	Al-Hujurat
4	An-Nisa'	27	An-Naml	50	Qaf
5	Al-Ma'idah	28	Al-Qasas	51	Adh-Dhariyat
6	Al-An'am	29	Al-'Ankabut	52	At-Tur
7	Al-A'raf	30	Ar-Rum	53	An-Najm
8	Al-Anfal	31	Luqman	54	Al-Qamar
9	At-Taubah	32	As-Sajdah	55	Ar-Rahman
10	Yunus	33	Al-Ahzab	56	Al-Waqi'ah
11	Hud	34	Saba'	57	Al-Hadid
12	Yusuf	35	Fatir	58	Al-Mujadalah
13	Ar-Ra'd	36	Ya Sin	59	Al-Hashr
14	Ibrahim	37	As-Saffat	60	Al-Mumtahanah
15	Al-Hijr	38	Sad	61	As-Saff
16	An-Nahl	39	Az-Zumar	62	Al-Jumu'ah
17	Al-Isra'	40	Al-Mu'min	63	Al-Munafiqun
18	Al-Kahf	41	Fussilat	64	At-Taghabun
19	Maryam	42	Ash-Shura	65	At-Talaq
20	Ta Ha	43	Az-Zukhruf	66	At-Tahrim
21	Al-Anbiya'	44	Ad-Dukhan	67	Al-Mulk
22	Al-Hajj	45	Al-Jathiyah	68	Al-Qalam
23	Al-Mu'minun	46	Al-Ahqaf	69	Al-Haqqah

70	Al-Ma'arij	85	Al-Buruj	100	Al-'Adiyat
71	Nuh	86	At-Tariq	101	Al-Qari'ah
72	Al-Jinn	87	Al-A'la	102	At-Takathur
73	Al-Muzzammil	88	Al-Ghashiyah	103	Al-'Asr
74	Al-Muddaththir	89	Al-Fajr	104	Al-Humazah
75	Al-Qiyamah	90	Al-Balad	105	Al-Fil
76	Al-Insan	91	Ash-Shams	106	Quraysh
77	Al-Mursalat	92	Al-Lail	107	Al-Ma'un
78	An-Naba'	93	Ad-Duha	108	Al-Kauthar
79	An-Nazi'at	94	Al-Inshirah	109	Al-Kafirun
80	'Abasa	95	At-Tin	110	An-Nasr
81	At-Takwir	96	Al-'Alaq	111	Al-Masad
82	Al-Infitar	97	Al-Qadr	112	Al-Ikhlas
83	Al-Mutaffifin	98	Al-Bayyinah	113	Al-Falaq
84	Al-Inshiqaq	99	Az-Zalzalah	114	An-Nas

Notes:

In this book, the Quran used for references is Tafsir Ar-Rahman, *Interpretation of the meaning of the Qur'an*, published by the Department of Islamic Development Malaysia, 2007.

Style of Citation of Quranic Verses in Texts

In citing the Quranic verses, the following approaches are used:

- In making a quotation, if the quotation is not for the whole verse, ellipsis "..." will be used.
- In making references to the Quran the style used e.g. Quran 9: 27 means Quran chapter 9 verse 27. For names of the chapter please refer to the list of *Surahs*.
- Refer to the stated chapter and verses for the full intention and meaning of the quoted verse.

Part 1

Shariah

Chapter 1

Shariah and Its Meaning in Islam

M. KAMAL Hassan

The term "Shariah" has been defined in various styles by different authors and researchers. The purpose of this chapter is to provide a reasonably clear and common understanding of the meaning of "Shariah."

The term "Shariah" has been defined as:

> "... water hole, drinking place; approach to a water hole; law", while *al-Shariah* (with the definite article) is defined as "... the revealed, or canonical law of Islam."[1]

Classical Arab lexicographers defined the term "Shariah" as:

> "Shariah and *mashra'ah* mean a watering place; a resort of drinkers [both men and beasts]; a place to which come to drink therefrom and to draw water, and into which they sometimes make their beasts to enter, to drink: but the term *mashra'ah*, or Shariah is not applied by the Arabs to any but [a watering place] such as is permanent, and apparent to the eye, like the water of rivers, not water from which one draws with the well rope; or a way to water because it is a way to the means of eternal life [shariah is] the religious law of God; consisting of such ordinances

---

[1] M. Kamal, H (2015). Understanding the Shariah and Its Place in the Muslim Communities: A Muslim-Malaysian Perspective. Keynote address at the International Symposium on Sharia in Asia-Pacific: Islam, Law and Politics, organized by the Religious Studies Programme, Victoria University of Wellington, New Zealand on August 25–27, 2015, quoting from Hans Weh's *Dictionary of Modern Written Arabic*.

as those of fasting, prayer and pilgrimage, and other acts of piety, or of obedience to God, or of duty to Him and to men: Shariah signifies also [A law, an ordinance, or a statute: and] a religion, or way of belief and practice in respect of religion and way of belief or conduct that is manifest and right in religion."[2]

Some of the aforementioned definitions are found in *Lisan al-Arab*, which defined Shariah linguistically as the place where people and animals come to drink water from it. It also says that Shariah and *Shir'ah* refer "to the religious duties that Allah has prescribed and commanded, such as fasting, prayer, pilgrimage, paying the *zakat* due and all the deeds of righteousness."[3]

In the Quran the term "Shariah" occurs in *Surah al-Jathiah* (Quran 45: 18):

"And now, We have set you (O Muhammad, and sent you) to help establish a way of religion (complete) constituting rules of religion; so you follow that way and do not yield to the desires of those who are ignorant (of the truth)."

M. Kamal (2015) reviewed the works of scholars of the Quran namely al-Tabari's *Tafsir Jami' al-Bayan*; Abd al-Rahman al-Sa'di's *al-Tafsir al-Karim al-Rahman fi Tafsir Kalam al-Manan*; Al-Shawkani's *Fath al-Qadir* and Mutawwali al-Sha'rawi's *Tafsir Khawatir* respectively and listed the meanings of the term "Shariah", namely:

- Shariah is the religious obligations (*al-fara'id*), and the divinely prescribed limits (*al-hudud*), and commandments (*al-amr*) and prohibitions (*al-nahy*) (which Allah has clearly laid down to be followed by human beings). There were those who equated *al-Shariah* with *al-Din* (the religion of man's complete submission to the One True God).
- Shariah is "A perfect way (*Shariah kamilah*) which calls to all that is good and prevents all that is evil. In following the way there is everlasting happiness (*al-sa'adah al-abadiyyah*), righteousness (*al-salah*) and wellbeing in this world and in the Hereafter (*al-falah*)."

---

[2]Lane, EW (1863). *Arabic-English Lexicon*, p. 1535. London: William and Norgate. www.Tyndalearchive.com/tabs/lane [Accessed August 16, 2015].
[3]Translated from Ibn Manzur (2003). *Lisan al-Arab*, Vol. 5(9), p. 40. Saudi Arabia: Dar Alamul Kutub.

- *al-Shariah* is "what Allah has ordained for His servants [to follow] with regard to the religion (*al-Din*)."
- The Shariah is the way that will lead to water which is "the origin of life."

In addition, al-Raghib al-Isfahani proposed the literal meaning of the word Shariah as "the way to a watering-place." He argued that the word Shariah is derived from the noun *Shar'* which means a clear open way for people to travel on. The Arabs have adopted the term Shariah to refer to the Divine Way (*al-Tariqah al-Ilahiyyah*), because according to some of them, the metaphor of the way to the watering-place implies that if "one were to enter into it in accordance with its true and validated nature, one would quench one's thirst [for the Truth] and become purified."[4]

Furthermore, Ibn Qayyim al-Jawziyyah explains the meaning of Shariah as follows: the edifice and foundation of the Shariah are divine authority (*al-hukm*) and the good welfare of the servants (of God) (*masalih al-'ibad*) in earthly existence and in the life to come. It is justice ('*adl*), all of it; mercy (*rahmah*), all of it; wisdom (*hikmah*), all of it; and (individual and public) welfare (*maslahah*), all of it. Therefore, any matter that draws away from justice to injustice (*al-jawr*), from mercy to its opposite, from welfare to causing corruption or perversion (*al-mafsadah*), from wisdom to foolishness (*al-'abath*), then it is not of the Shariah, even if it is included in it by allegorical interpretation (*al-ta'wil*) (Jasser, 2008a).

Viewing Shariah from the perspective of contemporary Islamic religion on traditional religious science, Yusuf al-Qaradawi explains the Shariah as follows: the Shariah is what Allah has ordained in the form of *ahkam* (commandments, injunctions, rules, regulations, laws), irrespective of whether they are primary (lit. "root", *asliyyah*) *ahkam* or subsidiary (lit. "branch", *far'iyyah*) *ahkam*. The Shariah is what Allah has made obligatory through His commands or prohibitions. This means that the Shariah is a totality of *ahkam*, some of which are commandments (*awamir*) to be carried out, prohibitions (*nawahi*) to be renounced. Some of them deal with forbidden matters

---

[4]Translated from al-Raghib, al-Isfahani (1999). *al-Mufradat fi Gharib al-Qur'an.* pp. 261–262. Beirut: Dar al-Ma'rifah.

(*muharramat*), some others deal with permissible matters (*halal*). So the whole of the Shariah contain these *ahkam*, and some of them are related to creedal matters (*al-'aqa'id*) as has been said by Sa'd al-Din al-Taftazani [1322–1390, a well-known Islamic polymath]:

> "Know that the rules and regulations (*ahkam*) of the Shariah, some of them are related to the proper ways of performing religious acts (*kaifiyyat al-'amal*) and they are called subsidiary (*far'iyyah*) and practical (*'amaliyyah*) aspects. And some of them are related to the fundamental principles of the religious acts (*asl al-'amal*) and they are called primary (*asliyyah*) and creedal (*i'tiqadiyyah*) aspects. The religious science that is related to the first aspect is called 'the knowledge of injunctions and rules, the science of jurisprudence (*'ilm al-fiqh*).' And the religious science that is related to the second aspect is the science of scholastic theology (*'ilm al-kalam*), or the science of Islamic monotheism (*'ilm al-tawhid*) or the creeds (*al-'aqa'id*)."[5]

Yusuf al-Qaradawi goes on to say that "Thus the Muslim scholars have agreed that all those (aspects) be called Shariah ordainments or rulings (*ahkam shar'iyyah*), except that when they refer, sometimes, to the "practical side" (*al-janib al-'amali*), they are *al-Shariah*, while the second is called *al-'Aqidah*. He supports this distinction by explaining how al-Azhar University had decided to establish two separate faculties, one called the Faculty of *Usul al-Din* (lit. "the roots of religion", meaning the science that deals with fundamental beliefs, theology, spirituality and related matters), the other called the Faculty of *al-Shariah* (injunctions, norms, rules, regulations and laws). Furthermore, he argued that the Shariah's main function (*mahammat al-Shariah*) is the same as the function of the mission of the Quran, namely:

> "... to take human beings out of darkness to light; the darkness of polytheism to absolute monotheism; the darkness of being astray to the divine guidance; the darkness of falsehood to the Truth; the darkness of ignorance to knowledge; darkness of chaos to order. Everything that brings order to life and guides mankind to the straight path is what the Quran brought, and these are the primary concerns and function of the Shariah, guiding human beings to what is good in it with regard to the religious and worldly aspects, to the life of discipline, of just balance."[6]

---

[5]M. Kamal, H (2015). Understanding the Shariah and Its Place in the Muslim Communities.
[6]Ibid.

Arising from the aforementioned definitions of the Shariah it is fair to conclude that the Quranic usage of the metaphor of "the path leading to water" in the term *al-Shariah*, is most fitting because like water, the God-given Way is absolutely necessary for the safety, security and the survival of human life. In other words, the Quran uses the most appropriate term in Arabic language to convey the supreme importance of following the divinely prescribed ordainments to ensure wellbeing in this world (*hasanah fi al-dunya*) and wellbeing in the Hereafter (*hasanah fi al-akhirah*).

Traditional Islamic religious scholars generally agree that the divinely prescribed *ahkam* (God's ordainments, judgments, rules, regulations and laws) which are meant to serve as a code of man's right conduct and action (*'amal*) as His servants, vicegerents and believers on earth are organically connected to, and inseparable from, the fundamental articles of faith (*iman*), the Creed of Absolute Monotheism (*'Aqidah* of *Tawhid*) which form the metaphysical presuppositions upon which the Islamic way of life, culture and civilization are to be constructed. Right faith (*iman*) and righteous action (*'amal salih*) as the inseparable twin preconditions for human wellbeing is reiterated in many places in the Quran.

However, on the issue of whether the Creed of Absolute Monotheism or fundamental Islamic beliefs are part and parcel of the Shariah, there is no uniformity of opinions among Muslim scholars on it. One group is of the opinion that Shariah embraces both the Islamic Creed as the fundamental beliefs or "root" (*al-asl*) of Islam and the divine rules, regulations and laws governing the practical aspects (*al-janib al-'amali*) as subsidiary component or "branch" (*al-far'*) of Islamic faith. The other group of Muslim scholars prefers to look at the Shariah as mainly dealing with prescribed religious obligations and practical dimension of the religion which include rules and regulations regarding matters of worship proper (*'ibadah*), such as the prayer (*salah*), fasting (*saum*), pilgrimage (*hajj*) and the *zakah*-tax, or compulsory charity as four of the five Pillars of Islam; marriage and divorce matters, the detailed rules and law of inheritance (*fara'id*), different types of commercial transactions which are not based on interest (*riba*), gambling (*maysir*) or uncertainties (*gharar*) but aimed at establishing a just economic system; and other matters related to human conduct in the spheres of good governance, criminal justice, international relations, war and peace.

It is important to clarify at this juncture that from the worldview of the Quran, the fundamental beliefs or *'Aqidah* of Islam, namely faith in the existence of the One True God, faith in His divine scriptures and all His human Messengers and faith in the existence of the Hereafter have remained the same throughout history. They are constant and unchangeable, but that part of the Shariah consisting of rules and regulations concerning forms of worship, specific economic transactions, special dietary rules, socio-cultural norms, etc., have been allowed by God to undergo changes and modifications to suit the changing circumstances.

The mission of the Shariah as Allah's ordained way towards mankind's welfare and wellbeing in this world and happiness and felicity in the Hereafter nevertheless, has indeed a broader scope of coverage and relevance insofar as it embraces not just the injunctions, rules and regulations covering all aspects of human life, but good moral values, ethical norms and moral laws. It is in the nature of the worldview of Islam that law, ethics and morality are basically intertwined and inseparable, unlike the positivistic theory of modern law which separates morality from law.

It should be clarified that "the ultimate objectives of the Shariah" (*Maqasid al-Shariah*) are, in fact, embedded in fundamental ethical values and norms, while the divine commandments and prohibitions covered by the main body of the Shariah are grounded in the divine moral principle of "enjoining that which is good and right (*al-ma'ruf*) and prohibiting that which is bad and wrong (*al-munkar*)" (Quran 3: 104, 113), this being a fundamental commandment from Allah to all believing Muslims and the Muslim community as a whole. The Sacred Law that the above commandment represents is at the same time the embodiment of His ethical values.

Thus, in following the Shariah Muslims are not merely obeying the Sacred Law. They are, at the same time, fulfilling ethical values or moral virtues such as:

- Obedience (*ta'ah*) to Allah, to His Final Messenger, and to the faithful leaders of the society;
- Justice (*'adl, qist*);
- Justly balanced moderation (*wasat*);
- Constant mindfulness and consciousness of divine pleasure and displeasure (*taqwa*);
- Moral-spiritual purification (*tazkiyah*); and
- Moral-spiritual excellence (*ihsan*), etc.

Moreover, in the Muslims' submission to the Shariah the Prophet of Allah (pbuh) serves as the best example to be emulated. For the Muslims the exemplar *par excellence* of the implementation of the Shariah in terms of the unity and inseparability of right action with right morality, of good deeds with *taqwa*, of just governance and leadership with ethical values of humility and egalitarianism, and of fundamental social-cultural reform, instilling a new moral discipline and undertaking societal restructuring with the moral values of tolerance, perseverance and compassion, was none other than the Final Messenger of Allah (pbuh) whose personality was declared by Allah Himself to have a "sublime moral nature" (Quran 68: 4).

The collection of authentic Prophetic traditions on good character (*husn al-khuluq*) as the most important criterion for the perfection of faith in Islam and for getting specially privileged status in the Hereafter amply testifies to the highest regard that Islam gives to the fulfilment of moral and ethical values.

The purpose of God raising Prophet Muhammad (pbuh), according to one tradition attributed to the Prophet Muhammad (pbuh), is the perfection of sublime morality ("I have been sent only to bring to perfection the virtues of moral conduct.").[7] The ultimate source for the Islamic vision of the inseparability of law

---

[7]M. Kamal, H (2015) quoting from *Musnad* of Ahmad Ibn Hanbal, *Mustadrak* of al-Hakim and *Kitab al-Adab* of al-Bukhari.

and morality is Allah Himself, who is — according to the creed of *Tawhid* — the Only Absolute Source and Criterion of what is absolutely right, lawful and good on the one hand, and what is absolutely wrong, unlawful and bad for His creatures (*makhluqat*) and servants (*'ibad*).

With its fundamental principles and values based on the Quran and the *Sunnah*, the Shariah as a whole is not a set of laws, although it is also normally translated as "Islamic law," but a combination of a comprehensive code of moral norms and ethical conduct as well as a body of divinely ordained religious Law. Thus the reward for good moral behavior or punishment for transgressing the rules of good conduct and moral behavior is not meted out in courts of this world but in the Hereafter, while the punishments for transgressing the penal laws (*hudud*), the laws of retributive justice or equal retaliation (*qisas*) and laws requiring discretionary judgment of the judge (*ta'zir*) are meted out in the courts of Islamic law on earth.

Therefore it is not surprising that the term Shariah is used by some Muslim scholars or intellectuals in the wider sense of the word, as the "revealed way of life," or "the heavenly part of Islamic law" (Jasser, 2008b, p. 57; Ramadhan, 1999, p. 28), while other scholars, particularly experts or specialists of Islamic jurisprudence (*fiqh*) or graduates of the faculties of Shariah in West Asia and North Africa normally use the term to refer to the more restricted sense of the concept, i.e. pertaining to the practical dimension (*al-janib al-'amali* as mentioned by Yusuf al-Qaradawi) which is covered by the collection of Allah's injunctions, rulings, regulations and laws as laid down in the Quran and elaborated in the Sunnah (normative tradition) of Prophet Muhammad (pbuh). Notwithstanding this latter tendency, it should not be forgotten that the whole scope and force of the Shariah include matters of religious worship, moral values, praiseworthy behavior and exemplary conduct, as well as matters related to the social, economic, political and international relations affairs, including the ethics of war and peace.

In the context of the five "Pillars of Islam,"[8] it is important to know that the most important pillar is the first Pillar which is a creedal matter regarding faith and conviction in the absolute oneness of God (*'Aqidah* of *Tawhid*) and oneness of His power and sovereignty — without any equal, or any partner of any kind as His associates or subsidiary deities, gods or supernatural beings, or anything partaking of His divinity, essence or attributes — over all aspects of human existence and life. It should be remembered that a person officially becomes a Muslim by pronouncing and believing in the first Pillar of Islam: the "Testimony of Faith" (*kalimat al-shahadah*) i.e. "I bear witness that there is no object of worship other than Allah, and I bear witness that Muhammad is the (Final) Messenger of Allah."

The five Pillars of Islam, obviously, do not come under the definition of law in the secular sense; they constitute the most basic and fundamental acts of religious worship (*'ibadah*) incumbent upon all Muslims to be performed to the best of their ability throughout their lives no matter where they may be. They are, nevertheless, considered as an important part of the Shariah of Islam.

Within the scope of Islamic jurisprudence (*fiqh*), the acts of religious worship (*'ibadah*) constitute the first part which always begins with matters of *'ibadah*. The second part deals with *mua-malat*, or types of lawful and unlawful commercial transactions, prohibitions in commercial and financial dealings, types of contracts and classifications of property and ownership. The third part deals with matters of *munakahat*, or matters related to Islamic marriage, its laws, divorce rules and regulations, followed by the subject of inheritance or law of succession with its detailed rules and regulations, guardianship, etc. The fourth part deals with *jinayah* or criminal offences spelled out in the Quran and the *Sunnah*, such as unlawful sexual intercourse, theft, robbery, false accusation of

---

[8]The Five Pillars of Islam are: (1) *Shahadah* (faith), (2) *Solat* (prayer), (3) *Zakat* (charity), (4) *Sawn* (fasting during the month of *Ramadhan*) and (5) *Hajj* (pilgrimage to Mecca).

unlawful sexual intercourse, consumption of intoxicating drinks or substances, and apostasy — grave offences which are considered to transgress the explicit "limits or boundaries" (lit. *hudud*) imposed by Allah to set the limits of human freedom.[9]

In conclusion, the Shariah as a whole is not a set of laws, although it is also normally translated as "Islamic law." Shariah is a combination of a comprehensive code of moral norms and ethical conduct as well as a body of divinely ordained religious law. The ultimate aims of the Shariah (*Maqasid al-Shariah*) are the preservation of:

- religion (*al-Din*)
- soul (*nafs*) or life
- mind or intellect (*'aql*)
- wealth or property (*mal*)
- progeny or offspring (*nasl*).

References

al-Raghib, al-Isfahani (1999). *al-Mufradat fi Gharib al-Qur'an.* pp. 261–262. Beirut: Dar al-Ma'rifah.
Ibn Manzur (2003). *Lisan al-Arab,* Vol. 5(9), p. 40. Saudi Arabia: Dar Alamul Kutub.
Jasser, A (2008a). *Fiqh al-Maqasid.* Herndon, VA: International Institute of Islamic Thought (IIIT).
Jasser, A (2008b). Maqasid Al-Shariah as Philosophy of Islamic Law. A System Approach. London: Internationanl Institute of Islamic Thought (IIIT). https://archive.org/stream/MufradatRaghibAlisfahani [Accessed August 16, 2015].

---

[9]*Hudud*: Limit or prohibition; pl. *hudud*. A punishment fixed in the Quran and *hadith* for crimes considered to be against the rights of God. The six crimes for which punishments are fixed are theft (amputation of the hand), illicit sexual relations (death by stoning or one hundred lashes), making unproven accusations of illicit sex (eighty lashes), drinking intoxicants (eighty lashes), apostasy (death or banishment) and highway robbery (death). Strict requirements for evidence (including eyewitnesses) have severely limited the application of *hudud* penalties. Punishment for all other crimes is left to the discretion of the court; these punishments are called *tazir*. With the exception of Saudi Arabia, *hudud* punishments are rarely applied, although recently fundamentalist ideologies have demanded the reintroduction of *hudud*, especially in Sudan, Iran and Afghanistan. http://www.oxfordislamicstudies.com/article/opr/t125/e757 [Accessed August 16, 2015].

M. Kamal, H (2015). Understanding the Shariah and Its Place in the Muslim Communities: A Muslim-Malaysian Perspective. Keynote address at the International Symposium on Sharia in Asia-Pacific: Islam, Law and Politics, organized by the Religious Studies Programme, Victoria University of Wellington, New Zealand on August 25–27, 2015.

Lane, EW (1863). *Arabic-English Lexicon*, p. 1535. London: William and Norgate. www.Tyndalearchive.com/tabs/lane [Accessed August 16, 2015].

Ramadhan, T (1999). *To Be a European Muslim*. Leicester: Islamic Foundation.

Chapter 2

Understanding of the Shariah in Regards to Construction

MOHAMAD AKRAM Laldin

The Meaning of Shariah

Islam as a religion is a complete way of life (*shumūl*) which encompasses the guidance of human affairs and needs. It demonstrates the finest way to conduct human life in private, economic, social, political, moral and spiritual affairs. Shariah regulates the practical aspects of Islam as a religion.

The word Shariah is derived from the root of *shin ra 'ayn* (شرع) which literally means, "the road to the watering place," "the straight path to be followed" and "the path which the believer has to tread in order to obtain guidance in this world and deliverance in the next." Shariah is "the way that directs man's life to the right path." As a technical term, the word Shariah is the law of Islam, all the different commandments of Allah (*s.w.t.*) to mankind (Nyazee, 2003).

A comprehensive definition of the word Shariah can be deduced from the different definitions given above as follows: it is the sum of Islamic teaching and system, which was revealed to Prophet Muhammad (pbuh), recorded in the Quran as well as deducible from the Prophet's divinely-guided lifestyle called the *Sunnah*.

The meaning implies the importance of Shariah in human life. Without Shariah, religion cannot exist in realized form within the heart and mind of its practitioner, and becomes like a soul without a body; it is out of this world without taking you there. If human wants to be in the world and proceed safely out of it, a Shariah is the vehicle to transport and to approach his Lord, which is used in contradistinction with *hawa* (passion). Therefore, the whole focus of Shariah is on humankind's *journey* towards intimacy with the Creator. The Quran tells us:

$$ثُمَّ جَعَلْنَٰكَ عَلَىٰ شَرِيعَةٍ مِّنَ ٱلْأَمْرِ فَٱتَّبِعْهَا وَلَا تَتَّبِعْ أَهْوَآءَ ٱلَّذِينَ لَا يَعْلَمُونَ ١٨$$

"And now, We have set you (O Muhammad, and sent you) to help establish a way of religion (complete) constituting rules of religion; so you follow that way and do not yield to the desires of those who are ignorant (of the truth)." (Quran 45: 18).

Components of Shariah

Shariah covers the total way of life that includes faith and practices, personal behavior, as well as legal and social transactions. All different commandments of Allah (*s.w.t.*) to mankind are part of Shariah. Each one of such commandments is called *hukm* (pl. *ahkām*) which describe rulings in the form of commands, prohibitions or permissibility on certain action from the Law Giver (*al-Shari'*).

Figure 2.1 presents the dimension of Shariah as a system of life that encompasses all aspects of the belief system (*al-ahkam al-I'tiqadiyyah*), the system of ethics and morals (*al-ahkam al-akhlāqiyyah*), and the practical rulings (*al-ahkām al-'amaliyyah*) governing human deeds that include man to God relationships (*'ibādah*) and man to man relationships (*muamalah*) (Mohamad Akram, 2008).

Shariah in this perspective covers the entire spectrum of life, including belief, morality, virtues and principles, guideline and framework of guidance on economic, political, social, cultural and civilizational matters that concern not only the Muslim community but all of humanity. Shariah is the body of Divine guidance, its structure, format and construct for human life.

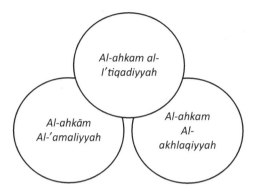

Figure 2.1. The Components of Shariah.

The Objectives of Shariah

Shariah aims at supporting human existence and providing the necessary principles, values and means to establish human wellbeing (*masālih al-'ibād*) in this worldly life and the hereafter (Quran 2: 30; 3: 191; 6: 165; 38: 27; 44: 38–39; 42: 13, 21; 45: 18; 67: 1–2).

The Shariah as a system of life has been revealed primarily to protect human wellbeing. All the Shariah's teachings, injunctions and prohibitions are related to the grand wisdom (*hikmah*) of blessing mankind and securing human interests. All the Shariah rules that contain obligations and duties bring benefit and prosperity and all its prohibitions prevent them from harm and hardship (Mohamad Akram, 2008).

The objectives of Shariah (termed as *Maqāṣid al-Shariah*) set the goals and objectives of Islam as a system of life that constitutes standards and criteria, values and guidance based on divine revelation (*wahy*) to be applied in practical life to solve human problems and guide the direction of human life in realizing benefits (*maṣlaḥah*) and avoiding harms (*mafsadah*). More narrowly framed, *Maqāṣid al-Shariah* is the underlying purpose upon which the Shariah is established (Mohamad Akram and Hafas, 2013).

Maqāṣid al-Shariah is defined by 'Allal al-Fassi (d.1974) as: "The overall objective of Islamic Law is to populate and civilize the earth and preserve the order of peaceful coexistence therein; to ensure the earth's ongoing wellbeing and usefulness through the piety of those who have been placed there as God's vicegerents; to ensure that people conduct themselves justly, with moral probity and with integrity in thought and action, and that they reform that which needs reform on earth, tap its resources, and plan for the good of all."

In the financial and construction sphere, *Maqāṣid al-Shariah* is considered the grand framework that provides guidelines and directions for ensuring the realization of *maslahah* (benefit) and the prevention of *mafsadah* (harm) in all financial and construction aspects such as contracts, processes, development, environment and society at large.

Muslim scholars divided *Maqāṣid al-Shariah* based on many considerations. For the construction industry, what concerns us is their division to *maqāsid* with regards to their strength and priority in application. Muslim scholars divided *Maqāṣid al-Shariah* into three main categories: the essentials (*darūriyyah*), the complementary (*hājiyyah*) and the embellishments (*tahsīniyyah*) (al-Ghazali, 1993).

First: *Darūriyyah* (necessities or essentials) is defined as interests of lives which people essentially depend upon, comprising the five aforementioned objectives of Shariah: religion (*dīn*), life (*nafs*), intellect (*'aql*), posterity (*nasl*) and wealth (*māl*). These are essentials serving as bases for the establishment of welfare in this world and the hereafter. If they are ignored then coherence and order cannot be established and (chaos and disorder) *fasād* shall prevail in this world, and there will be obvious loss in the hereafter. Some scholars argued that though the five *darūriyyah* are essential for human welfare, necessities are not confined to these five *maqāsid*, hence they proposed additional *darūriyyah* such as equality, freedom and the protection of the environment.

Second: *Hājiyyah* (need or complementary) are interests that supplement the essential interests. It refers to interests whose neglect leads to hardship but not to total disruption of the normal order of life. In other words, these interests, other than the five essentials, are needed in order to alleviate hardship, so that life may be free from distress and predicament. It also acts as provisions that aim at removing hardships and/or facilitating life. An example is seen in the sphere of economic transactions where the Shariah validated certain contracts such as the *bay' salām* (forward sale) and also that of (*ijārah*) lease and hire because of the people's need for them, notwithstanding a certain anomaly that is present in both.

Third: *Tahsīniyyah* (embellishments) refer to interests whose realization leads to refinement and perfection in the customs and conduct of people at all levels of achievement. For example, the Shariah encourages charity to those in need beyond the level of the obligatory *zakat*. In customary matters and relations among people, the Shariah encourages gentleness, pleasant speech and manner, and fair dealing. Other examples include permission to use beautiful, comfortable things; to eat delicious food; to wear fine clothing and so on.

Sources of Islamic Law

Sources of Islamic law in Islamic jurisprudence is termed as *dalīl* (pl. *adillah*) which literally means guide. Al-Āmidi (1991) mentioned *dalīl* as "a valid examination of which yields transmitted information" (*mā yumkin al-tawassul bi sahīh al-nazr fī hi ilā matlub khabari*). The Quran, in this sense, is a *dalīl* for the *ahkām* of Allah, because it leads us to the *khitāb* (desire) of Allah that contains the *ahkām* related to the acts of the subjects. Therefore, the sources of Islamic law are called *al-adillah al-Shar'iyyah* or *al-masādir al-Shar'iyyah*.

In general, Shariah rulings are derived from the revealed sources, namely the Quran and the *Sunnah*. Nevertheless, it is also undeniable that there are other sources that are recognized by the majority of jurists and are important in deriving Shariah rulings (*ahkām*). In the following we will classify the sources of Islamic law, see their legitimacy and elaborate on the example of their application in Islamic capital market.

Shariah recognizes two sources of knowledge, the sources based on revelation (*wahy*) which are called *al-adilah al-naqliyyah* and the sources based on intellectual reasoning (*'aql*) which are called *al-adillah al-'aqliyyah*. The Quran and the *Sunnah* are the sources that originate from the text of revelation which is called textual sources (*al-naṣṣ*), and other sources based on the power of reasoning are called non-textual sources (*al-ra'y*).

The sources of Shariah are also classified into primary and secondary sources. The primary sources of Shariah are the Quran, *Sunnah, ijma'* and *Qiyās*. The secondary sources of Shariah are *istihsān* (juristic preferences), *istishāb* (presumptions of continuity), *al-masālih al-mursalah* (consideration of public interest), *sad al-dhara'i* (blocking the means to the evil) and *'urf* (customs).

The scholars can derive various laws, including laws related to the construction industry from those sources. Having various channels of knowledge and multiple sources for law derivation characterizes Islamic law as a universal law, flexible and dynamic in application and adaptation to the contexts, circumstances and places.

Shariah Fundamentals and the Construction Industry

Riba

Riba is derived from the verb *raba* that literally means "to grow" or "expand" or "increase" or "inflate" or "excess." Any excess in quantity, addition, an increase of a thing over and above its original size or amount is called *riba*.

Nevertheless, that is not what is meant in the technical definition of *riba*. It does not mean every increase or growth is prohibited in Islam. Trade (*al-bay'*) for example, is considered as legal means for wealth accumulation and growth. Al-Jurjani in his *al-Ta'rifāt* defined *riba* as "any increment from the exchange which is conditioned upon one of the contracting parties."

Based on its sources, *riba* could happen in two places: loan (*Riba al-duyūn*) and sale transaction (*Riba al-buyū'*):

Riba al-Duyūn

Riba al-duyūn is *riba* that arise in loan transactions because of the time given to repay as well as for further extensions or delays in repayment. Some scholars divide it further into two types:

(1) **Riba al-Qard** is any increment in borrowing or lending money whether in kind or cash, over and above the principal amount, as a condition stipulated or agreed between the parties. It is prohibited by Shariah, to put condition (*mashrūṭ*) to take any addition or benefit from the loan given based on the Islamic legal maxim:

كل قرض جر نفعا فهو الربا

"Every loan which brings a benefit is riba."[1]

(2) **Riba al-Jahiliyyah** is an extra charge in addition to the debt principal as a result of deferment in paying the debt to the creditor. The people of Arab at that time practice this form of *riba* whereby on the day of debt maturity the debtor will ask the creditor, *anzirni azidka* (give me more time, I will pay more). This *riba* is prohibited in the *hadith* during Rasulullāh (pbuh) performed *Hajj al-Wadā'*:

وإن ربا الجاهلية موضوع ولكن لكم رؤوس أموالكم لا تظلمون ولا تظلمون وقضى الله أنه لا ربا. وإن أول ربا أبدأ به عمي العباس بن عبد المطلب

"All interest obligation shall henceforth be waived. Your capital, however, is yours to keep. You will neither inflict nor suffer any inequity. Allah (*s.w.t.*) has Judged that there shall be no interest, and that all the interest due to my uncle, Abbas ibn Abdul Muttalib shall henceforth be waived."

Riba al-Buyū'

Riba al-buyū' occurs from trading or exchange transactions in which a commodity is exchanged for the same commodity from the *ribāwi* commodities in unequal amount and/or delay of the delivery of one of the commodities. It refers to *riba* which results from non-compliance to the two conditions stipulated in the *hadith* mentioned

---

[1]The origin of this Islamic legal maxim is a *hadith* narrated by al-Baihaqi, al-Baghawi and others, however the *hadith* is weak as stated by scholars of *hadith*, nevertheless, scholars consider this legal maxim in their *fiqhi* examinations and Shariah rulings.

below in transactions involving the six items enumerated whereby the commodities should be exchanged in equal counter-value and on the spot.

Based on its cause, *riba* could occur because of time factor (i.e. deferment, *nasā'*) and quantity factor (i.e. not equal, *tafādul*). Based on this *riba* is classified into *riba al-fadl* and *riba al-nasi'ah*:

(1) **Riba al-Fadl** is "extra addition in exchange of two commodities (*al-ziyādah fi ahad al-'iwadain*)."
(2) **Riba al-Nasi'ah** is "deferment in possession of two commodities to certain time (*ta'khīr fi qabd ahad al-'iwadaini ilā ajal mu'ayyan*)."

This classification is based on the *hadith* reported by Ubadah bin Samit that the Prophet (pbuh) said:

«سَمِعْتُ رَسُولَ اللَّهِ - صلَّى اللَّهُ عَلَيْهِ وَسَلَّمَ - يَنْهَى عَنْ بَيْعِ الذَّهَبِ بِالذَّهَبِ، وَالْفِضَّةِ بِالْفِضَّةِ، وَالْبُرِّ بِالْبُرِّ،

وَالشَّعِيرِ بِالشَّعِيرِ، وَالتَّمْرِ بِالتَّمْرِ، وَالْمِلْحِ بِالْمِلْحِ إِلَّا سَوَاءً بِسَوَاءٍ عَيْنًا بِعَيْنٍ، فَمَنْ زَادَ أَوِ ازْدَادَ فَقَدْ أَرْبَى»

"Gold for gold, silver for silver, wheat for wheat, barley for barley, dates for dates and salt for salt, like for like, equal for equal and hand to hand, if the commodities differ, then you may sell as you wish provided that the exchange is hand to hand" (Sahih Muslim, Book 10, Hadith 3853).

This *hadīth* indicates two main criteria to constitute *riba*, the first is deferment of the time of exchange and secondly, the difference of counter-values in the exchange of two similar *ribāwi* (the commodities specified in the *hadith*) items. Table 2.1 shows the ruling of *riba* transactions, based on the *hadith* will be as follows:

• In relation to the exchange of commodities with the same category and types such as gold and gold, or dates and dates the condition is:
  ○ *Al-tamāthul* which is *equality* in measurement either weight or count
  ○ *Al-hulul* which is *on the spot* (no delay or postponement) in delivery
  ○ *Al-taqābud* which is complete possession in the *majlis al-'aqd* before the parties involved are separating or leaving the session of contract.

Table 2.1.   *Riba* Categories and Rules.

Category	Type	Exchange	Quantity
Same category	Same type (e.g. Gold with Gold; Wheat with Wheat)	Spot exchange	Equal in quantity regardless of quality
Same category	Different type (e.g. Gold with Silver; Wheat with Rice; Salt with Dates	Spot exchange	Inequality is permitted
Different category	Different type (e.g. Gold with Wheat; Ringgit Malaysia (RM) with Dates)	Delayed is permitted	Inequality is permitted
Ribāwi items and non-*ribāwi* items (e.g. RM with vehicles; USD with furniture)		Delayed is permitted	Inequality is permitted
Between two non-*ribāwi* items (Cloth with bird; bricks with sands)		Delayed is permitted	Inequality is permitted

- In relation to the exchange of commodities with the same category but different types (such as gold and silver), the condition is:
  - *Al-hulul*: the contract must be *on the spot* (no delay or postponement) in delivery
  - *Al-taqābud*: complete possession in the *majlis al-'aqd*.
- In relation to the exchange of commodity with different categories and types there is no condition.

Gharar

Gharar is among the prohibited elements in commercial transactions. Nevertheless, unlike *riba*, *gharar* prohibition is not absolute in the sense that some degree of *gharar* is tolerable. Only excessive *gharar* (*gharar fāhish*) where uncontrollable risk leads to speculation and gambling must be avoided.

Gharar also implies uncertainty and deceit which lead to dispute and cause injustice to any of the parties. Shariah is very concerned with that and therefore it requires the parties to the contract to be as specific as possible so that their agreements may not give rise to future disputes.

Gharar literally means risk (*alkhatr*). It also means reduction (*al-nuqsān*) and what is pleasant on the surface but unpleasant underneath. *Gharar* generally refers to uncertainty, ambiguity, high risks, unknown and even ignorance.

Gharar, according to Al-Sarakhsi (1994), technically means "Anything that the end result is hidden or the risk is equally uncommon, whether it exists or not." In Mu'jam ISRA (2010) *gharar* is defined as "Something for which the probability of getting it and not getting it are about the same; some said: something whose acquisition is uncertain and its true nature and quantity are unknown."

Both definitions imply *gharar* as something which is hidden and uncertain by both parties of the contract either on the nature of the object of contract or the outcome arising from the contract.

The prohibition of *gharar* can be deduced from the saying of Allah (*s.w.t.*):

يَـٰٓأَيُّهَا ٱلَّذِينَ ءَامَنُوا۟ لَا تَأْكُلُوٓا۟ أَمْوَٰلَكُم بَيْنَكُم
بِٱلْبَٰطِلِ إِلَّآ أَن تَكُونَ تِجَٰرَةً عَن تَرَاضٍ مِّنكُمْ وَلَا تَقْتُلُوٓا۟
أَنفُسَكُمْ إِنَّ ٱللَّهَ كَانَ بِكُمْ رَحِيمًا ﴿٢٩﴾

"O believers! Do not consume (use) your wealth among yourselves illegally (such as by means of cheating, gambling and others of illegal nature), but rather trade with it by mutual consent. And do not kill one another. Indeed Allah is Most Merciful to you." (Quran 4: 29).

Gharar is also forbidden by Islam based on the *hadith* reported by Ibn 'Umar (R.A):

نهى رسول الله صلى الله عليه وسلم عن بيع الغرر

"The Messenger of Allah forbade sales that involve gharar." (Sunan Ibn Majah, Vol. 3, Book 12, Hadith 2194).

The *'illah* (cause) for *gharar* prohibition is *jahālah* (ignorance which leads to dispute and hatred). *Gharar* is prohibited to protect the wellbeing of both parties and ensure a satisfactory outcome of a contract. It is forbidden to ensure full consent and satisfaction of the parties in a contract. Without full consent, a contract may not be valid. Full consent can only be achieved through certainty, full knowledge, full disclosure and transparency. The prohibition of *gharar* also shows that the spirit of Islam in contracts or transactions is that of transparency, accuracy and disclosure of all necessary information. Hiding information, deceit and fraud is not allowed. Likewise, having no knowledge in pursuing a transaction is not desired.

Gharar in a transaction is located either in the contract itself (*sīghah al-'aqd*) or in the subject-matter of contract (*mahal al-'aqd*). It is characterized as the object and price of contract is in-existence (*ma'dūm*) unknown (*majhul*) or incapable to be delivered (*ma'jūz 'an al-taslīm*).

The existence of *gharar* in a contract might expose the parties involved in the contract to excessive risk due to uncertainty of its outcome. At the end, it might typically lead to enmity and dispute. Avoiding *gharar* will result in clearing the ambiguity and reducing risk. Therefore, it is in the best interests of both buyer and seller to have complete knowledge and access to the product before the transaction, specifically in the object of sale of what is being bought, sold and at what price.

Gharar is divided into three types:

(1) *Gharar yasīr* (light *gharar*) where uncertainty is slight or trivial and hence might be easily tolerated by both parties. Examples given by Muslim jurists include the foundation of a property and the embedded material of a cloth or sofa.

(2) *Gharar mutawasit* (intermediate *gharar*) where Muslim jurists may hold different views based on the *ijtihad* in either associating it to *Gharar yasīr* (light *gharar*) or *Gharar fāhish* (excessive uncertainty).

(3) *Gharar fāhish* (excessive *gharar*) is uncertainty that affects the core element of the subject matter such as selling a bird in the sky.

Excessive uncertainty is prohibited in various *hadiths* of the Prophet (pbuh). We can summarize the causes of prohibition as follows:

(1) **Uncertainty of existence and future outcome of a transaction**, such as the sale of an item that may not exist or is not in the possession of one of the parties; the sale of birds in flight, fish not yet caught, stray (runaway) animal, sale of an unborn calf in its mother's womb or sale of unripened fruit on a tree.
(2) **Uncertainty in possession and ownership**, such as the sale of goods whereby there is no ownership or incomplete ownership such as selling something owned by another person without authority, or selling something which is not certain in terms of its possibility to be delivered.
(3) **Inadequacy and inaccuracy of information**, such as in type, shape, quantity, weight and sum, time delivery, price value and payment method.
(4) **Undue complexity of the contract**, such as interdependent and conditional contracts as a result of combining two sales in one contract; two sales are linked jointly; or conditional sales so the fulfilment of one sale is conditional upon the fulfilment of the other; or the sale is conditional on external event. Pure games of chance (*al-qimār* and *al-maysir*), sales which is done based on games or chance.

Muslim jurists gave some exception to the general prohibition of *Gharar fāhish*, where excessive uncertainty is tolerated. Among these exceptions are:

- There is a public need for the transaction or contract (consideration of *maslahah*), for example, *salām* and *istisnā'*. Although the object of sale does not exist, Rasulullāh (pbuh) allows that transaction-provided specification on the price, product and time delivery is explicated.
- The contract is unilateral or charitable (*al-tabarru'āt*) such as gift or bequest. Unspecified amount of gift, donation or rebate to be given voluntarily by one person to the other is tolerated. This is

because those contracts are charitable in nature and no counter-value is expected in return.

Maysir (including Speculation)

Maysir means any activities which involve betting, whereby the winner will take the entire bet and the loser will lose his bet. *Maysir* is also defined as "acquiring wealth by the means of merely putting properties at risk."

Maysir refers to avoidable uncertainty as it involves taking risk that is created in the contract itself (contractual risk, as opposed to trade risk) in the sense that the parties involved are taking unnecessary uncertainties which is not part of everyday life. *Maysir* is a game of chance which operates on the basis of speculation. Nevertheless, Shariah does not prohibit general types of speculation. Only speculation which is akin to gambling, i.e. gaining something by chance rather than productive effort is prohibited.

The contract of *maysir* will not bring benefit to both parties, instead, it operates on a win-lose situation whereby only one party will gain and the other will lose. In other words, the bottom line of the *maysir* game is "either of us who wins take the other's property."

Maysir to some extent is similar to *gharar* (uncertainty). Both share similar features of uncertainty over gain and loss. Likewise, the unknown outcome from a transaction that contains *maysir* or *gharar* might bring either benefit or harm to either parties. *Maysir* and *gharar* can be interrelated, where there are elements of *gharar*, the contracting parties are essentially exposing themselves to risk-taking and gambling (*mukhātara wa qimār*). For example in *bay' al-mulāmasah* (the touch-and-throw sales), the buyer has to touch the material to find out the quality of the product. In case of the buyer not knowing or touching the material, this is considered as *bay' al-gharar* which contains the element of gambling (*maysir*).

Nevertheless, *maysir* is different from *gharar*. *Maysir* is played for its own sake and is often played as a game whereas *gharar* proceeds over sales and contracts. *Gharar* is usually not the purpose of a contract but incidental to it whereas *maysir* is the purpose of the game, which has no other subject matter, or purpose other than winning and beating one's opponent in order to take his property.

Maysir is prohibited because of the uncertainty involved, the possibility of deception as well as the enmity that may arise between the parties if the expectation of any of them is not met. In the Quran *maysir* is strictly prohibited in the following verse:

$$\text{يَٰٓأَيُّهَا ٱلَّذِينَ ءَامَنُوٓا۟ إِنَّمَا ٱلۡخَمۡرُ وَٱلۡمَيۡسِرُ وَٱلۡأَنصَابُ وَٱلۡأَزۡلَٰمُ رِجۡسٞ مِّنۡ عَمَلِ ٱلشَّيۡطَٰنِ فَٱجۡتَنِبُوهُ لَعَلَّكُمۡ تُفۡلِحُونَ (٩٠)}$$

"O believers! Wine and gambling, idols, and divining arrows are (all of them) abomination divised by Satan, Avoid them, so that you may prosper." (Quran 5: 90).

Manipulation

Shariah prohibits all type of deceptions. *Ihtikār* (hoarding), *najsh* (artificial price hiking), *tadlīs* (concealment of a defect) and other prohibited acts are considered elements or techniques of manipulation which are prohibited in Shariah.

Ihtikār is an act of manipulation by purchasing essential commodities, and hiding them from the market to cause an increase in price because of the artificial dearth of supply in the market.

Shariah prohibits hoarding and considers it a market manipulative practice that involves the creation of artificial shortages in the supply of a specific needed commodity by hoarding up large quantities thereof in warehouses to keep supply shorter than demand in order to profit at the expense of others (often consumers) by badly affecting the very sustenance of people and distorting the market price.

The Prophet Muhammad (pbuh) has prohibited hoarding on several occasions. The Prophet (pbuh) said:

$$\text{مَن احْتكَر حكرةً يريدُ أنْ يُغالي بهَا على المسلمين فهو خاطئٌ}$$

"Whoever practiced the hoarding necessities of life with the intention to create artificial increase thereby (profiteering) is a sinner." (Musnad Ahmad Ibn Hanbal).

Najsh literally means concealment or a practice in which hunters used to rouse and chase the game for the sole purpose of trapping or snaring it. Technically, Imām Al-Shafi'i defined *bay' al-najsh* as "the sale wherein a person bids-up the price of a commodity with no intention of buying it, only to induce others to buy it for more than they would have otherwise." This trick is forbidden by the *hadith*:

عن ابن عمر رضي الله عنهما قال: نهى رسول الله صلى الله عليه وسلم عن النَّجْش

On the authority of 'Abdullah ibn 'Umar (R.A): "The Prophet (pbuh) forbade al-najsh." (Bukhari and Muslim, Book 18, Hadith 1581).

Thus, jurists condemned the behavior of both the seller (who is part of the conspiracy) as well as the potential buyer who drums-up the goods, and the latter was labelled a sinner. This is the seller attempting to sell a commodity for a price higher than its fair value by conspiring with fake bidders to ensnare a buyer.

Tadlīs literally means concealing, deceiving or hiding. Technically, *tadlīs* is an act of intentional concealment of the goods' defects in a sale transaction. *Tadlīs* is prohibited by the Shariah, the evidence of prohibition is based on the *hadith* of the Prophet (pbuh) that says:

"عن أبي هُرَيْرَةَ أنّ رَسُولَ اللّهِ صلَّى اللّهُ عَلَيْهِ وَسَلَّمَ قال: " لا تُصَرُّوا الإِبِلَ وَالْغَنَمَ، فمَنْ ابْتَاعَهَا بَعْدَ ذَلِكَ

"فهُوَ بخَيْرِ النّظرَيْن بَعْدَ أن يحلبها، إن رضيا أمسكها، وإن سخطا رَدَّهَا، وصَاعًا مِنْ تَمْر

"Do not forcefully keep the milk in the udders of camels and sheep, and if one buys it thus, then he has the option after milking it, he may keep it or return it together with a container of dates as compensation for the milk." (Sahih Muslim, Book 21, Hadith 15).

Jurists agree in general that the option given to the buyer if the defect was concealed by the seller is based on deception; however, they consider the contract as valid.

Shariah Contracts Related to Construction

Before discussing the Shariah contracts related to construction, it will be better to understand the general concept of contract in Shariah. The word contract or *'aqd* in Arabic is mentioned in the Quran and the Quranic injunction "*Aufu bi al-Uqud*": "Fulfil Your obligations" (5: 1) is the fundamental principal which governs the sanctity of all contracts, whether private, public, civil or commercial.

The literal meaning of "*'Aqd*" (Plural: *Uqud*) is to tie or bind. As a technical term *'aqd* can be defined as "The obligation which is the result of an offer given by one party and the acceptance given by the other party, in a way where its legal effect is expressed on the things contracted upon." Some jurists defined the term *'aqd* in a wider sense to cover every covenant drawn by a person (unilateral) which include bequest (*al-wassiyyah*), charitable endowment (*wkaf*) and oath (*yamin*).

The above definition of *'aqd* indicates its two important elements: the connection of the word of the contracting parties and the binding of the two intentions expressed through words or otherwise. In addition, these connections imply responsibilities of both parties to execute the agreement in the contract.

Fundamentals of Contract

There are three pillars (*arkan*) of contract in which these pillars must be in existence without which will render any contract to be void and unusable:

(1) The expression of offer and acceptance. This is to indicate the consent and intention to enter and execute the contract (*al-ijab wa al-qabul*).
(2) The performers of contract or the parties involved in the contract.
(3) The object of contract, i.e. the subject matter and the countervalue or consideration.

The following discussion will examine each and every one of the pillars of contract.

The Expression of Offer and Acceptance (Al-Ijab Wa Al-Qabul)

Ijab is the expression of the first party of the contract to express the offer and *qabul* is the expression of the other party to indicate the acceptance. The expression must be clear and understandable and can be done either by word, action, in writing or by signs. Shariah emphasizes the importance of expressing the will (*ridha*) of the contracting parties. In modern day transactions, it can be done over the phone by e-mail, short messaging services (sms) and other means that serve the purpose of indicating the will of the parties involved, provided that the parties can be ascertained as genuine parties performing the offer and acceptance through these modes of communication. It is advisable to have something written as proof of the parties involved in the contract.

Performer of Contract

The performer of contract (the contracting parties) must be eligible to enter and execute the contract and bear all its consequences. Those given this right must fulfil the following conditions:

(1) Reach the age of puberty (*baligh*).
(2) Person with sound mind (*rushd*).

The Object of Contract/Subject Matter

The objects of contract are the items or things on which the contract is concluded. This includes:

(1) Moveable or immovable assets such as house, land, car, etc.
(2) Non-assets such as human beings in the marriage contract.
(3) Benefits in the lease contract or jobs in the case of an employer.

There are several conditions concerning the subject matter of contracts as follows:

(1) Existence of the object — The object must be in existence at the time of the contract if it is an actual thing. The existence can be at the place where the contract is concluded or at some other place. Some kinds of contract do not require the existence of the objects i.e. contract of *salam* (forward buying) or *istisna'* (manufacturing contract).
(2) Legality — The object must be legal from the Shariah point of view and it is a commodity capable of being traded.

(3) Certainty of delivery — The object must be able to be delivered and not something beyond the means to be delivered, such as the existence of an object whose whereabouts is uncertain.

(4) Precise determination of the object — The exact specifications of the object must be clearly determined so as to avoid any future dispute. If it is a car the type, model, year of manufacture and other relevant details should be determined.

The above pillars of contract must be fulfilled in any contract related to construction and failure to fulfill any of the conditions will render the contract null and void. It must be emphasized that consent is very important in a contract and any element that will lead to the deficiency in consent might affect the validity of the contract. For example if there is any element of force for someone to enter into the contract, it will render the contract as null and void. Similarly, any ambiguity in the description and disclosure of a construction agreement might lead to the contract being null and void.

Conditions of Contract

The parties in a construction contract can agree to any condition as long as these conditions do not contravene the objectives of the said contract. For instance, a selling and buying contract with the condition that the buyer will have limited use of the subject matter after the contract is concluded. This condition contravenes the objective of the contract which is the complete transfer of ownership including the usufruct (usage) of the subject matter.

Selected Types of Contracts

There are various classifications of contracts in Islamic law. The main types of contract and their brief descriptions are listed below.

Contracts of Exchange ('Uqud Al-Mu'awadat)

The primary concern of this contract is trading as well as selling and buying activities inclusive of their sub-divisions such as cash sale, deferred payment sale, deferred delivery sale, sale on order, sale of currency, auction sale and other forms of sale. The details of some commonly used types of sale are as follows:

Bai Muajjal (Deferred Payment Sale)

A contract involving the sale of goods on a deferred payment basis. The bank or provider of capital buys the goods (assets) on behalf of the business owner. The bank then sells the goods to the client at an agreed price, which will include a mark-up since the bank needs to make a profit. The business owner can pay the total balance at an agreed future date or make installments over a pre-agreed period.

Murabahah Sale

Contract of sale of a commodity at the cost price plus a known profit. Some scholars illustrate it as an agreement where the vendor specifies the cost price of the commodity that he bought and sells it at a declared profit. The profit may be declared in exact value such as when the vendor says, "I bought this commodity for ten Ringgit and selling it at a profit of two Ringgit." The profit may also be declared in proportion, such as the vendor saying, "I am selling this commodity at the profit of one Ringgit for every ten Ringgit that I spend for its cost."

Bai al-Salam

This term refers to advance payment for goods which are to be delivered later. Normally, no sale can be effected unless the goods are in existence at the time of the bargain. But this type of sale forms an exception to the general rule provided the goods are defined and the date of delivery is fixed. The objects of this type of sale are mainly tangible things but exclude gold or silver as these are regarded as having monetary values. Barring these, *bai al-salam* covers almost all things which are capable of being definitely described as to quantity, quality and workmanship. One of the conditions of this type of contract is advance payment; the parties cannot reserve their option of rescinding it but the option of revoking it on account of a defect in the subject matter is allowed. It is also applied to a mode of financing adopted by Islamic banks. It is usually applied in the agricultural sector where the bank advances money for various inputs to receive a share in the crop, which the bank sells in the market.

Bai al-Istisna'

Istisna' is a contract whereby a party undertakes to produce a specific thing which is possible to be made according to certain agreed-upon specifications at a determined price and for a fixed date of delivery. This undertaking of production includes any process of manufacturing, construction, assembling or packaging. In *Istisna'*, the work is not conditioned to be accomplished by the undertaking party and this work or part of it can be done by others under his control and responsibility.

Bai Istijrar

A contract between a supplier and a client whereby the supplier supplies a particular item on an ongoing basis on an agreed mode of payment until they terminate the contract. It is also applied between a wholesaler and a retailer for the supply of a number of agreed items.

Bai al-Wafa'

This is a contract with the condition that when the seller pays back the price of the assets sold, the buyer returns the assets to the seller. It is a sale in form but a pledge in substance. Some scholars claim that it is a security contract in the form of sale based on the fact that both parties to the contract have the right to claim the exchanged items. This means that the seller has the right to claim the goods he has sold by paying the buyer the full price of the goods sold. It is called *wafa'* (fulfilling the obligation) because of the obligation to fulfil the condition in the contract which is returning the items sold when the seller claims it by paying back the price.

Bai Muzayadah

This is the sale of an asset in public through the process of bidding among potential buyers and the asset is sold to the highest bidder.

Contract of Lease (Ijarah)

Contract pertaining to the utilization of usufruct e.g. contract of lease (*ijarah*). *Ijarah* is a contract under which a party leases out an asset for a rental fee. The duration of the lease and rental fees are agreed in advance. Ownership of the equipment remains in the hands of the lessor. Another form of *ijarah* is *ijarah wa iqtina'* (lease purchase). It is similar to *ijarah*, except that the lessee is committed to purchase the equipment at the end of the rental period. It is pre-agreed that at the end of the lease period the lessee will purchase the assets at an agreed price from the lessor, with rental fees paid to date, forming part of the price.

Contracts of Partnership (Shirkah)

It includes the following types of contracts:

Musharakah

Musharakah (partnership) which means a partnership where profits are shared as per an agreed ratio whereas the losses are shared in proportion to the capital/investment of each partner. In *musharakah*, all partners to a business undertaking contribute funds and have the right, but not the obligation, to exercise executive powers in that project.

Mudharabah

Mudharabah (profit and loss sharing) is a form of commercial contract whereby one of the contracting parties act as capital provider (termed as *rabb al-mal*) while the other manages the enterprise (termed as *mudharib*). If there is loss, the provider of capital bears the financial loss while the worker/manager loses his labor. If there is profit, both parties share it in proportions agreed upon at the time of the contract.

Contract of Security Such as Pledges (Rahn) and Suretyship (Kafalah)

Rahn

Rahn means pledging an asset or property as a security for a debt or a right of claim, the payment in full of which is permitted from the sale of the asset or property in the event of default by the debtor.

Kafalah (Suretyship)

Legally this means the pledge given by the guarantor or the surety (*al-kafil*) to a creditor on behalf of the principal debtor to secure that the guaranteed (*al-makful bih*) will be present at a definite place to pay his debt or fine.

Contract of Gratuity Which Includes Hibah, Al-Ibra' and Wakaf

Hibah

Hibah (gift) is giving one's wealth to others without the expectation of any replacement or exchange with the transferring effect on the ownership. Therefore, once a *hibah* is executed, the giver cannot take it back. Like *wakaf* or other charity, *hibah* is the transfer of property from the giver to the recipient during his lifetime.

Al-Ibra'

Al-ibra' is giving up of a right. In a commercial transaction a creditor gives up part or all of his right to a debtor usually for early settlement of the debt.

Wakaf

Wakaf (endowment) which means a form of gift in which the corpus is detained and the usufruct is set free for the confined benefit of certain philanthropy for the sake of Allah. Detention of the corpus implies preventing it from being inherited, sold, gifted, mortgaged, rented, lent, etc. The utilization of *wakaf* property shall also comply with the purpose mentioned by the *waqif* (donor) and without any pecuniary return.

Contracts of Trusts (Al-Amanat)

Contracts of trusts (*al-amanat*), such as safekeeping (*wadi'ah*) which means empowerment to someone for keeping the owners' wealth explicitly or implicitly.

Contracts to Do a Specified Task Such As Commission (Ju'Alah) and Agency (Wakalah)

Ju'alah

Ju'alah is a unilateral contract promising a reward for the accomplishment of a specified task.

Wakalah

Wakalah is a contract of appointment of an agent where a person appoints another as his agent to act on his behalf.

Sulh

Contracts to resolve conflicts or disputes such as arbitration and mediation. *Sulh* is a form of contract legally binding on both the individual and community levels and the purpose of *sulh* is to end conflict and hostility among different parties so that they may conduct their relationships in peace and amity. In *muamalah*, the concern is settlements of financial disputes among people.

Conclusion

Shariah consists of different commandments of Allah (*s.w.t.*) to mankind and is the way that directs man's life to the right path. Each one of such commandments is called *hukm* (pl. *aḥkām*) which describe rulings in the form of commands, prohibitions or permissibility on certain action from the Law Giver (*al-Shari'*). Shariah covers the total way of life that includes faith and practices, personal behavior, as well as legal and social transactions.

The sources of Shariah rulings are derived from the primary sources: Quran, *Sunnah*, *Ijmā'* and *Qiyās*; and secondary sources: *istihsān* (juristic preferences), *istishāb* (presumptions of continuity), *al-masālih al-mursalah* (consideration of public interest), *sad al-dhara'i* (blocking the means to the evil) and *'urf* (customs). The sources provide principles, objectives as well as practical rulings on human life.

The objective of Shariah (*maqāṣid al-Shariah*) aims at supporting human existence and provides the necessary principles, values and means to establish human wellbeing (*masālih al-'ibād*) in this worldly life and the hereafter.

Shariah plays a role in the construction industry by setting sources and principles that guide the construction industry, sets the methodology of *ijtihād* (deriving practical rulings from the sources), sets norms and parameters that guide the structuring and implementing of the construction industry as well as operations, products, finance and anything related to it.

References

Al-Āmidi, in (1991). *Al-Ihkām fī Usūl al-Ahkām*. Beirut, Damaskus: Al-Maktab al-Islāmi.

Al-Ghazali, AHM (1993). *Al-Mustasfa min 'ilm al-usul*. 'Abd al-Salam 'Abd al-Safi, M. (Ed.). Beirut: Dar al-Kutub al-'Ilmiyyah.

'Allal al-Fasi (1985). *Al-Madkhal li Dirasa al-nazariyya al-'Amma li al-Fiqh al-Islami wa Muqaranatuhu bi al-Fiqh al-Ajnabi*. Rabat: Mu'assasat.

Al-Sarakhsi (1994). *Al-fusūl fī al-Usūl*. Kuwait: Wizāratu al-Awqāf Wa al-Shu'ūn al-Islāmiyyah.

International Shariah Research Academy for Islamic Finance (ISRA) (2010). Mu'jam ISRA li al-Mustalahāt al-Māliyah al-Islāmiyyah. Kuala Lumpur: ISRA.

Mohammad Akram, L and F Hafas (2013). Developing Islamic Finance in the Framework of Maqasid al-Shari'ah: Understanding the ends (maqasid) and the means (wasa'il). *International Journal of Islamic and Middle Eastern Finance and Management*, 6(4): 278–289.

Mohammad Akram, L (2008). *Fundamentals and Practices in Islamic Finance*. Kuala Lumpur: ISRA.

Nyazee, IAK (2003). *Islamic Juridsprudence*. Kuala Lumpur: The Other Press.

Chapter 3

Promoting Efficiency in Construction Practices: Lessons from Shariah

AINUL JARIA Maidin

Introduction

Construction activities are an integral part of a country as continued development to meet the needs of the growing urban and rural communities is unavoidable.

Overall, it is undeniable that construction industries everywhere are facing problems and challenges. They include poorly implemented methods of designing, building and maintenance of buildings and infrastructures; 3Ds (dangerous, dirty and demanding) poor performance and the lack of ethics and integrity. However, in developing countries, the problems and challenges are present alongside a general situation of socio-economic stress, chronic resource shortages, institutional weaknesses and a general inability to deal with the pertinent issues. There is also evidence that they have become greater in extent and severity in recent years.

This chapter seeks to recommend the adoption of Shariah principles and values, which can help promote and reduce the weaknesses plaguing the construction industries of the developing countries besides promoting efficient development of the practices in the construction industry.

Lessons from the Shariah

In Islam development activities must be goal-oriented and value-realizing. They involve confident and all-pervading participation of men and directed towards maximization of human wellbeing in all aspects and building the strength of the *ummah* so as to enable men to discharge their roles as Allah's (*s.w.t*) vicegerents on earth. Development encompasses moral, spiritual and material development of the individual and society at large leading to maximum socio-economic welfare and the ultimate good of mankind (Zubair, 2006).

Shariah principles guiding development activities provide a direction for Muslims to build some of the finest cities, in terms of their physical amenities, and their incredible ability to organize citizens into a homogenous and integrated society (Beg, 1983). Most Islamic cities were built on top of existing layers of local or regional history without affecting the existing civilization (Ibrahim, 1988). The pillars of Islam require the Muslims to develop a highly specialized and organized social structure through which they can demonstrate their faith. Prophet Muhammad (pbuh) himself was instrumental in establishing the City of Madinah, which played an important role in providing sustainable living to the people along with conservation of natural resources in propagating Islam (Abdulac Samir, 1984). The Quran and *Sunnah* do not favor the nomadic Bedouin life (Abdulac Samir, 1984). Early Muslim towns such as those in the *Maghreb* like Al-Fustat, Tunis and Rabat were built to preach the religion of Islam and played the role of Citadel of Faith (Rabah, 2003). They all were compatible with the environment and its processes (Rabah, 2003).

The Muslim world may have suffered some drawbacks in terms of development in comparison with the industrial nations. There are always two sides to a coin. Development will not necessarily bring positive changes nor can we say it is harmful. Certain developments such as clean water, environmental standard and good sewerage management are essentials that can be adversely affected by poorly planned construction activities. Harmful effects of development such as indiscriminate land development, cutting of matured trees, clearing hillyland, slopes and development near seashores have had negative impact on the Earth's ecosystem.

The Quran and *Sunnah* speak of God's design for creation and humanity's responsibility for preserving it. The objective of Shariah guidelines is aimed at preservation of religion (*al-deen*) in order to achieve spiritual contentment, achieve physical and mental (*al-nafs*) wellbeing of man and his descents (*al-nasab*); mental strength (*al-aql*); and the appropriate way of dealing with all worldly property and wealth (*al mal*) (Moustapha, 1986). This is to achieve the ultimate aim of creating a healthy and virtuous individual, family, society (*ummah*) and eventually an ideal world (Moustapha, 1986). It was proclaimed in the Quran in *Surah Ali Imran*: 138 to the effect that, "This is a declaration for mankind, a guidance and instruction to those who fear God" thus Islam has provided answers to the problems faced by mankind.

Promoting a better quality of life for the *ummah* includes protection of every living being, and preservation of natural resources through the development of a just, moral, creative, spiritual, economically vibrant, caring, diverse and cohesive society characterized by high productivity. The Shariah principles prescribes conservation of natural resources so that (i) they will not be depleted by the present generation's use and deprive the future generation; (ii) the built environment (*al-imarah*) is in harmony with the natural environment and the relationship between the two is designed to be one of balance and mutual enhancement; (iii) preservation of environmental quality by reducing processes that degrade or pollute the environment and to prevent development that are detrimental to human health or that diminish the quality of life; (iv) prevent any development that affects social equality and unequal distribution of the earth's resources; (v) effective participation (*shura*) by all stakeholders in the planning decision-making and development control processes (Besim Salim Hakim, 1986).

The goal of Islam is one of sustainable development for the sake of achieving *mardatillah* (pleasure of Allah (*s.w.t*)) without sacrificing the right to promote reformation and transformation in accordance with the needs of the society and advancement of technology. In the context of development activities, Islam requires the following elements to be given consideration (Kurshed, 1980):

(1) *Tawhid* — God's unity and sovereignty;
(2) *Rububiyyah* — divine arrangements for nourishment, sustenance and directing things towards their perfection;
(3) *Khilafah* — man's role as God's vicegerent on earth; and
(4) *Tazkiyah* — purification plus growth.

All development activities must be carried out in accordance with the rules and controlled by the government. The Quran pronounces that it was revealed to reform human conditions. Despite accepting the pre-Islamic customs and practices of Muslims, Islam has rejected customs which were harmful to society and development to prevent harm and oppressiveness. Governments are given the rights to exercise their power of reasoning (*ijtihad*) to develop new rules and regulations based on the Quran and *Sunnah*. Rules are an important tool for promoting the orderly administration of any civilised society. Most Muslim nations have not managed to fully develop their legal system based on the Quran and *Sunnah* except in areas of personal law, finance and criminal law. They continued to implement the Western colonial law, regulations and administrative system even after achieving independence.

There are many biased accusations of the Muslims being anti-development, oppressive and backward. These allegations are baseless and do not hold water. Islam has never been against development and improvement. This is proven through a series of verses which begins with the responsibility of Muslims to seek knowledge. Allah imposed the obligation for Muslims to learn in the Quran, again and again. In fact, the first verse that was revealed to Prophet Muhammad (pbuh) is Quran 96: 1 which means "Read! (O Muhammad) in the name of your Lord, Who has created (all that exists)."

Allah (*s.w.t*) encourages Muslims to acquire knowledge for the sake of development and improvement of *ummah* which will subsequently lead to the expansion of Islam. Quran 11: 37, for instance, Allah (*s.w.t*) instilled the knowledge of the construction of an ark to Prophet Noah (a.s) through divine inspiration. The ark was big and strongly built which construction would be impossible without divine intervention. Such sophistication and technology, even during that era, is glaring proof that Islam values development and promotes growth in tandem with development activities.

The Quran in 14: 32 provides that:

> "It is Allah Who created the heavens and the earth, and sends down water from the sky with which He produces fruits for your sustenance. And it is He Who made the ships subject to you so that they may sail the ocean in your service by His leave. And the rivers also He has made them subject to you (so that you can get benefit from them)."

Referring to the practice of Prophet Muhammad (pbuh), Yusuf al-Qaradhawy (2009) commented that whenever the Prophet (pbuh) is faced with matters related to the end result of an agriculture technique which he did not quite like, he would just say to the Companion, "You know better in your earthly matter" and said that, "whatever the Muslims deem to be good is good in the eyes of Allah."

Manner of Deriving Law Under Islamic Legal System

Islamic law is derived from two distinct sources: Quran and *Sunnah*. It is a body of legal rulings, judgments and opinions that have been collected over the course of the centuries. On any point of law, one will find many conflicting opinions about what the law of God requires or mandates. The Islamic legal tradition is expressed in works that deal with jurisprudential theory and legal maxims, legal opinions (*fatwa*), adjudications in actual cases and encyclopaedic volumes that note down the positive rulings of law (*ahkām*). Islamic law covers a broad array of topics ranging from rituals to criminal law, family law, commercial and transaction laws, international law, constitutional law and others. The scholars of Islam (*ulama*) have generally considered *Rahmah* (Mercy) to be the all-pervasive objective of the Shariah and have, to all intents and purposes, used it synonymously with *maslahah* to promote benefit to man's daily affairs (Mohammad Hashim, 1999, p.1). Similarly, Schact (n.d) observes: "...the underlying tendency of the Quranic legislation was to favor the underprivileged; it started with enunciating ethical principles...." This feature of Quranic legislation was preserved by Islamic law, and the purely legal attitude, which attaches legal consequences to relevant acts, is often superseded by the tendency to impose ethical standards on the believer (Schact, 1950).

What is customarily referred to as Islamic law is actually separated into two distinct categories that is Shariah and *fiqh*. Shariah is the eternal, immutable and unchanging law, or Way of truth and justice, as to how Allah (*s.w.t*) wants men to live by. Thus, men must strive and struggle to realize Shariah law to the best of their abilities. *Fiqh* on the other hand is man-made law in the context of attempting to reach and fulfil the requirements stipulated by Allah (*s.w.t*). *Fiqh* is subject to error, alterable and contingent.

The moral and ethical objectives of the Quran play a central and pivotal role in the process of legal analysis. The point of the legal analysis is not to unthinkingly and blindly implement a set of technical rules, but to seek the ultimate objectives of the Quran. All Quranic laws reinforce and promote moral and ethical objectives and it is the duty of Muslims to apply themselves intellectually in order to comprehend and fulfil these objectives.

The specific rulings of the Quran came in response to particular problems that confronted the Muslim community at the time of the Prophet (pbuh). The particular and specific rules set out in the Quran are not objectives in themselves. These rulings are contingent on particular historical circumstances that might or might not exist in the modern age. At the time these rulings were revealed they were sought to achieve particular moral objectives such as justice, equity, equality, mercy, compassion, benevolence and so on. It is essential that Muslims study the moral objectives of the Quran, and treat the specific rulings as demonstrative examples of how Muslims should attempt to realize and achieve the Quranic morality in their lives.

Shariah requires man to live in accordance with the teachings of the Quran and *Sunnah* to achieve divine grace (*barakah*) for the hereafter. Thus, all development activities planned by man in this world must emphasize on piety (*taqwa*), manifest the oneness of

God (*tawhid*), peace and most of all manifestation of humility and humble oneself to the Creator's greatness in all aspects of man's life (Amini Amir, 1997; Sayyed Hossein, 1968; Ismawi, 1996). The act of a Muslim revolves very much around his intention for which he is held accountable.

Imam Idris Al Azhar in the process of building the town of Fez said his prayers, before beginning construction, as follows (de Montequin, 1980):

> "Oh my Lord, You know that I don't intend by building this city to gain pride or to show off, nor do I intend hypocrisy or reputation or arrogance, but I want you to be worshipped in it; Your laws, limits and the principles of your Quran and the guidance of Your Prophet to be upheld in it, as long as this world exists. Almighty, help its dwellers to do righteousness and guide them to fulfil that. Almighty prevent from them the evil of their enemies, bestow your bounties upon them and protect them from the sword of evil. You are able to do all things."

Thus, for a Muslim, the intention (*niyyah*) must be right to arrive at the true goal. Prophet Muhammad (pbuh) said: "Everything is according to the niyyah." The intention is closely related to the heart (*al-qalbun*) of the Muslim. The professional involved in construction needs to ensure that his activities are organized strategically to comply with the principles of Shariah to achieve pleasure of God (*mardatillah*). A building designer must not plan with the intention solely to derive profit at the expense of the people, and the environment, as this can cause lack of safety of the people and degrade the environment.

The contemporary system promotes the idea that construction is a purposeful activity, i.e. a deliberate, conscious striving towards the purposeful end of improving "standards of life" irrespective of how vague this goal may be and this is blindly accepted by the general public (Faludi, 1973). Construction professionals often want to provide progress indicators to convince the public of their success (Baster, 1972). The public only reacts when a building fails in total or partly on the basis that the contractor had failed to deliver certain promised material progress or anticipated standards of life.

The construction process guided by the Shariah is basically different from the contemporary construction system as it guides man to develop by maintaining the harmonious balance between man and his Creator, man and man, and man and the natural environment. The preceding paragraphs will discuss the Shariah principles that can be adopted to promote best practices in the construction industry.

Guidelines in Formulating Law on Construction and Maintenance of Built Environment

Many have thought that Islamic law or the Shariah focuses only on criminal punishments or sanctions like *hudud* and *qisas*. Some might restrict Shariah to personal laws such as matrimonial, inheritance and custody of children. Indeed, many will be surprised to know that Islam is very concerned with the regulations pertaining to construction.

Islam is a complete religion with a complete set of rules and guidance on all aspects of life. Interestingly, the Federal Court of Malaysia in the case of *Meor Atiqurrahman* [2006] 4 CLJ 1, in its judgment acknowledged the fact that Islam "is not just a religion but a complete way of life."[1]

Therefore, every single rule promulgated must be in line with the Shariah. Not only that, every single act of human being is judged and will either carry legal accountability or spiritual accountability. Tracing the concept of buildings in Islam shows that every rule pertaining to the development or maintenance of the buildings must be parallel to the principles enunciated in Islam (Mohammad Hashim, 2009).

The Quran expressly provides in numerous places and a variety of contexts, of the rationale, purpose and benefit of its laws so much so that its text becomes characteristically goal-oriented, explains

---

[1]The Court of Appeal in applying "the integral part of the religion" test relied on a number of Indian authorities like *The Commissioner, Hindu Religious Endowments, Madras, v. Sri Lakshmindra Thirtha Swamiar* AIR [1954] SC 282, *Sardar Syedna Taher Saifuddin Saheb v. State of Bombay* AIR [1962] SC 853, *Javed v State of Haryana* (AIR) [2003] SC 3057 and *Commissioner of Police v. Acharya Jagadishwaranada Avadhuta* [2004] 2 LRI 39. The Court also referred to the judgment of the Malaysian Supreme Court in *Hajjah Halimatussadiah bte Hj. Kamaruddin v. Public Services Commission Malaysia & Anor* [1994] 3 MLJ 61 where the Supreme Court applied the same test.

cause and benefit in the affirmative sense, evil conduct and crime is reprimanded and made punishable, in order to prevent injustice, corruption and prejudice. The underlying theme in virtually all the broad spectrum of the *ahkam* is realization of benefit (*maslahah*) which is regarded as the core principle of *maqasid*. For justice is also a *maslahah* and so is educating the individual (*tahdhib al-fard*) (Jasser, 2008).

Islam acknowledges the importance of the State developing rules to ensure that the State's affairs are administered and the rights of the people are protected. Allah (*s.w.t*) commands the authority to administer in accordance with the rules prescribed in the Quran and man is required to develop, implement and enforce the rules for the benefit of man and the Universe.

"Truly, Allah commands you to hand back your trusts to their (rightful) owners, and (Allah commands you) whenever you have to judge between people, to pass judgment upon men with fairness. Indeed, most excellent is that which Allah exhorts you to do. Truly Allah is All-Hearing, All-Seeing." (Quran 4: 58).

The authority has vast powers to make and impose rules for the organized administration of the State. However, with the given power, Islam imposes a strict restriction and enforces heavy responsibilities on the authority, which failure of compliance will lead to Allah's (*s.w.t*) punishments.

Allah (*s.w.t*) proclaimed in Quran 5: 45:

"... And those who do not judge in accordance with what Allah has revealed, they are the evildoers."

In Quran 5: 44, it is proclaimed that:

"... For those who do not judge in accordance with what Allah has revealed, they are disbelievers."

Allah (*s.w.t*) cautioned the decision-makers in the Quran 5: 49 that:

"And so pronounce judgement among them in accordance with what Allah has revealed and do not be led by their desires. Take heed lest they turn you away from a part of that which Allah has revealed to you. If they turn away (by refusing to accept Allah's injunctions) know that it is Allah's Wish to punish them for some of their sins; and many of the people are wrongdoers."

This is further endorsed by the *hadith* narrated by Abu Dharr:

"I said to the Prophet (pbuh), 'Messenger of Allah, will you not appoint me to a public office?' He stoked my shoulder with his hand and said, 'Abu Dharr, thou art weak and authority is a trust, and on the Day of Judgment it is a cause of humiliation and repentance except for one who fulfils his obligations and (properly) discharges the duties attendant thereon.' (Sahih Muslim, Book 20, Hadith 4491).

Thus, it is clear that the authority has the right to regulate in accordance with Shariah, which includes the laws and regulations on construction. Having looked at the nature of the responsibility entrusted upon the authority, it is only fair that the authority be given full cooperation and obedience with regards to the laws and regulations they promulgate. It is a duty of a Muslim to adhere to the laws set by the authority. Allah (*s.w.t*) proclaimed in Quran 4: 59:

"O believers! Obey Allah, and obey The Messenger and those entrusted with authority among you. Should you differ in anything among yourselves, refer it to (the Book of) Allah (the Quran) and (the tradition of the) Messenger, if you truly believe in Allah and the Last Day. This will be best (for you) and best in the end."

In addition, a *hadith* narrated by Abu Huraira in which Prophet Muhammad (pbuh) said:

"It is obligatory for you to listen to the ruler and obey him in adversity and prosperity, in pleasure and displeasure, and even when another person is given (rather undue) preference over you." (Sahih Muslim, Book 20, Hadith No. 4524).

However, the laws and regulations decreed by the authority must always be in compliance with Allah's (*s.w.t*) rulings. In a *hadith* narrated by Ibn 'Umar, the Prophet (pbuh) said:

"It is obligatory upon a Muslim that he should listen (to the ruler appointed over him) and obey him whether he likes it or not, except that he is ordered to do a sinful thing. If he is ordered to do a sinful act, a Muslim should neither listen to him nor should he obey his orders." (Sahih Muslim, Book 20, Hadith No. 4533).

Developing Islamic Principles of Contract to Enhance Compliance

In legal usage, Muslim jurists have defined an *"aqd"* contract as "The obligation and engagement of two contracting parties with reference to a particular matter. It expresses the combination of offer and acceptance. In the conclusion of contract, both the offer and the acceptance are interrelated in a legal manner, the result of which is seen in their mutual relationship (Mejelle, 2000, pp. 103–104). Based on the definition in the Mejelle (2000), the requirement of the existence of "two contracting parties" and the need for the "combination of the offer and the acceptance" are clearly stated. This definition means that by far, the most common and consistent use of the word *aqd* was in relation to transactions that were concluded by offer and acceptance.

The construction industry can adopt either the "price in advance" or *al-Istisna* contract or the cost reimbursement or *al-Ijarah* contract. The former poses a high risk to contractors while the latter places a high risk on the client.

An interestingly large number of projects in the Malaysian construction industry today practice the *al-Istisna* contract or "price in advance." In a typical scenario for public projects, the contractor can earn about 10% profit (depending on the type of negotiation with the client), allowing them to further sub-contract the project to sub-contractors. The practice of multilayer sub-contracting resulted in the existence of management contractors (with no specialty), high dependency on domestic sub-contractors, issues on corruption, and delayed and unfair payments to the bottom of the supply chain, among others (Hamid, 1969). Operating businesses or projects with minimum profit will adversely affect business survivability. These issues seem to pose a major effect on the operations of contractors and sub-contractors and are strongly linked to the high rate of business failure in the construction industry.

In comparison, an *al-Ijarah* contract or cost reimbursement allows contractors to be paid only based on their work, with an acceptable allowance for profit and attendance in the case of direct sub-labor (Hamid, 1969). For large projects, the management contract system allows the involvement of professional advisors to manage all suppliers and expert sub-contractors (not necessarily domestic contractors), who are hired for a particular project and who do the work based on their specialties. This type of contract does not allow the hiring of an inexpert management contractor and the utilization of multilayered sub-contracting.

Recognition of Shariah Principles in International Contracts

Problems are posed to the Muslims seeking to apply Shariah law in regulating the contractual transactions either between Muslims or between Muslims and non-Muslims. The courts have discussed whether Shariah is sufficiently certain in order to be applied as the proper law of a contract in the cases of *Musawi v. Re International (UK) and Others* [2007] EWHC 2981 (Ch) (*"Musawi"*), *Beximco Pharmaceuticals Ltd & Others v. Shamil Bank of Bahrain EC* [2004] EWCA Civ 19 (*"Beximco"*) and *Halpern v Halpern & Anor* [2007] EWCA Civ 291 (*"Halpern"*).

In the case of *Musawi*, an Iraqi citizen and an Iranian family concluded a series of agreements to jointly acquire, develop and own a piece of land adjoining the Wembley Stadium in London. Subsequently, the parties disagreed as to their respective rights and entered into an arbitration agreement providing that the dispute should be settled in accordance with "Islamic legal standards." They also appointed an Ayatollah "as arbitrator and Islamic legal judge." The arbitrator issued an award in favor of the Iraqi party, who sought enforcement in the High Court of Justice. The Court ruled in favor of the plaintiff. The Court referred to Dicey, Morris & Collins, *The Conflicts of Laws* (14th edition), affirmed that section 46(1)(b) of the UK 1996 Arbitration Act "allows parties the freedom to apply a set of rules or principles which do not in themselves constitute a legal system" and that "such a choice may include a non-national set of legal principles (such as the 1994 UNIDROIT Principles of International Commercial Contracts) or, more broadly, general principles of commercial law or the *lex mercatoria*." Consequently, the Court held that the arbitration agreement validly selected Shia Shariah law as the law applicable to the substance of the dispute and decided in favor of the enforcement of the arbitral award.

In the case of *Beximco*, the English Court of Appeal held that a choice of the principles of Shariah law was not a choice of law of a country for the purposes of the Rome Convention. The judge held that English law was the governing law because there could not be two separate systems of law governing the contracts. The words used were intended to reflect the Islamic religious principles according to which the bank held itself out as doing business, rather than a system of law. The parties had not chosen Shariah law as the governing law because: (1) it was not the law of a country and there was no provision for the application of a non-national system of law such as Shariah law. Further, it was highly improbable that the parties had intended that an English secular court should determine any dispute as to the nature or application of such controversial religious principles. In interpreting the governing law clauses, the court should lean against a construction which would defeat the commercial purpose of the agreements. There could not be two governing laws in respect of the agreements. The Rome Convention 1980, scheduled to the Contracts (Applicable Law) Act 1990, only contemplated and sanctioned the choice of the law of a country; (2) although it was possible to incorporate provisions of foreign law as terms of a contract, the general reference in the agreements to principles of Shariah law did not identify any specific aspects of Shariah law intended to be incorporated into the contracts. The reference to Shariah law was repugnant to the choice of English law and could not sensibly be given effect to. The judge was right that the words were to be read as a reference to the fact that the bank held itself out as conducting its affairs according to the Shariah principles. The judge was also correct that a common mistake as to the legal consequences of the agreements would not give rise to a defence to the claims on the guarantees, because the borrower's sole interest was to obtain advances of funds and they were indifferent to the form of the agreements required by the bank or the impact of Shariah law on their validity.

In *Halpern*, the Court of Appeal considered an application for summary judgment to enforce an alleged compromise of a Beth Din[2] arbitration. The application raised two interesting issues: the extent to which parties are able, under the provisions of the Rome Convention, to apply *Halakha* (Jewish law), which is not the law of any "country," and the extent to which rescission of a contract for duress at common law requires the claimant to make counter-restitution. In accordance with the orthodox view, the Court of Appeal held that a choice of Jewish law would not be effective under the Rome Convention. It declined to determine the issues relating to counter-restitution, holding that these questions should not be addressed until the court had heard all the evidence.

One of the challenges at present in using Islamic law of contract is that it needs to be established that Islamic Law is not an unsophisticated, obscure and defective system, and that it is a basic element of the Islamic society and can be explored further with the initiatives of the Muslim world. It is important to be aware that in Muslim countries like Saudi Arabia and other parts of the Middle East, the law and principles of Shariah relating to contracts are not codified. Certain general principles of Shariah are applied in order to assist in the interpretation of construction and all types of contracts. In the absence of a codified law and judicial precedents, in providing a legal opinion, parties often rely upon their own interpretation of the general principles applied in Saudi Arabia and based on their knowledge of legal practice applied to similar cases or in similar circumstances in the past. Nevertheless, under the general principles of Islamic law, the overriding principle contained is the maxim *"the contract is the law of the parties"* (Tarek Badawy, 2012). This has become accepted, meaning that, in general, the parties to a contract are free to agree to the terms of their

---

[2] A Beth Din is a rabbinical court of Judaism. In ancient times, it was the building block of the legal system in the Biblical Land of Israel.

choice, provided that these terms are not at odds with established Islamic principles and incorporated into the contract between the parties.

Provisions Preventing Destruction to Natural Resources during Construction

Islam is a religion which supports development but at the same time demands moderation. A life in moderation (*wasatiah*) is enjoined on Muslims. Totally challenging development and being moderate are two different things. Moderation can be seen even in the way Allah (*s.w.t*) explained the creation of creatures in Quran 15: 19:

> "And We have spread out the earth, and placed on it immovable mountains, and We have caused life of every kind to grow on it in a balanced manner."

Islam is against extremism or excessiveness, as proclaimed by Allah (*s.w.t*) in Quran in 31: 19 that:

> "Rather be moderate in your pace, and lower your voice (when you talk). The ugliest of all voices is the braying of the ass."

Islam avoids any act of transgressions or excessiveness as Allah (*s.w.t*) despises transgressions. In one of the earliest revelations, Allah (*s.w.t*) proclaims in Quran 5: 2 that:

> "O believer! Do not violate the sanctity of the rites of Allah, or the sacred months, or the animals offered (for sacrifice in Makkah), or the garlands of the animals, or those who go to the Sacred House to seek the bounty of their Lord (through commerce) and His pleasure (by performing his Hajj at the Holy Place). Once your pilgrimage is ended, you shall be free to go hunting. And do not let your hatred for those who has debarred you from the Holy Mosque lead you to transgress. Help one another in furthering virtue and God-consciousness, and not in what is wicked and sinful. And remain conscious of Allah, for He is stern in retribution (to those who disobeyed His orders)."

Applying the same principle, it is just to suggest that there must be some limitations imposed on the construction industry. First and foremost, the construction of buildings cannot lead to excessive destruction to the environment and natural resources. The prohibition of destroying the environment is mentioned more than once in the Quran. Allah (*s.w.t*) in Quran 2: 205 reminded the believers that:

> "No sooner does he leave you (after getting what he wants) then he hastens to commit corruption in the land, destroying crops and offspring (of cattle and mankind), and Allah do not love mischief."

That Allah (*s.w.t*) despised destruction of the environment can be seen in His prohibition of destroying the surroundings and trees unnecessarily during war. Allah (*s.w.t*) in Quran 2: 190 proclaimed that:

> "And fight for the sake of (establishing and protecting the religion of) Allah those who fight against you, but do not commit aggression (by initiating the fight), for Allah does not love aggressors."

The methodology of deriving the rulings or *Usul fiqh* is crucial in Islam. Under the rule of *Usul fiqh*, there is a methodical maxim which carries the meaning that "whenever there is any contradiction between 'prohibition' and 'allowance', then the 'prohibition' shall prevail" (Mawil Izzi, 2004). Interestingly, in providing examples on the application of the above maxim, the construction of roads, among others, was discussed. Accordingly, if a secondary public service, such as the building of the road or factory, could lead to adverse effects on public health or the environment, then such a road or factory should not be sacrificed. Thus, it is apparent that construction of roads must be within the allowed parameters such as observing the contour and not to cut through hills unnecessarily if it can be avoided. Strict observations are obligatory to preserve the environment (Fazlun, 2002).

If one examines closely, these concepts of moderation and limitations set in construction of buildings and infrastructure despite acknowledging the importance of development to human life, one can see that Islam observes the "carrying capacity" of the earth which is the core element of sustainable development (Portney, 2003). As defined by the United Nations' World Commission on Environment and Development (WCED) which is better known as the Brundtland Commission, "Sustainable Development" is "development that meets the needs of the present without compromising the ability of the future generations to meet their own needs" which conceptual definition is also shared with the National Commission on the Environment (NCE) of the United States. The NCE further added that "sustainable development is premised on living within the earth's means" (Portney, 2003).

Incorporate Duty to Maintain Buildings

It is unmistakable that the government or any other authorities has the power to make laws and regulations pertaining to construction and maintenance of buildings. Islam views the responsibility imposed on the authority as an important aspect that must be enforced.

Although there are no specific Quranic text or *hadith* on building maintenance, one may by inference conclude that the maintenance shall be within the responsibility of the authority. The authority must ensure that the roads connecting people are in good condition so that the people can go to places of worship, or conduct businesses, or travel to seek knowledge in comfort. Allah instructs the Muslims to walk on the earth in search for Allah's *rizk* (sustenance). Allah (*s.w.t*) proclaimed in Quran 62: 10 that:

> "Then when the prayer is finished, then disperse through the land (to carry on with your various duties) and go in quest of Allah's bounty and remember Allah always (under all circumstances), so that you may prosper (in this world and the Hereafter)."

Quran 67: 15 provided that:

> "It is He Who has subdued the earth for you to use it; so walk about its regions and eat of His provisions; and (remember) to Him (you shall return after) the resurrection, (so you must make the best use of His bounties and fear His punishment)."

These needs must be provided and protected by the authority. They must assure that the ways towards good are enjoined. Allah (*s.w.t*) proclaimed in Quran 22: 41, that:

> "That is, they are those (believers) who, if We firmly establish them on earth, will attend to their prayers and pay the alms-tax, enjoin the doing of what is right and forbid the doing of what is wrong. And (remember) with Allah rests the final outcome of all affairs."

A strong indication to support the above idea can be seen in the seriousness of Islam in preventing handicaps caused by accident. Rispler-Chaim (2007) underlined how Islam strives to reduce the cases of disability. He is of the opinion that Islam adopts preventive measures rather than curing. This is based on the principles of *hifz al-masalih* and *dar' al-mafasid* aimed at promoting preservation of good causes and preventing bad causes. This is indeed very true as accidents can at times cause irreparable harm and permanent disabilities.

Owing to the concept of accountability and seeking Allah's (*s.w.t*) pleasure, taking care or maintaining the built environment is a collective responsibility in Islam. Local authority does this due to its legal duty and *amanah* but the public, at the individual level, does so to discharge his religious or spiritual duty owed to Allah (*s.w.t*).

Allocation for Payment of Compensation for Injuries Arising from Poorly Constructed Buildings and Environment

Payment of compensation to users for injuries or damages suffered due to poor building conditions is an important aspect that needs to be incorporated into any proposed legislation. As analyzed from a case-based approach, damages incurred by any member of the public which is caused by the failure on the part of the authority to execute its duty properly must be compensated by that particular authority. This is in line with the Shariah. This can easily be deduced from the strictness of Islam in its effort to eliminate all possible causes of disability. Islam in this aspect has adopted preventive measures and ruled that every law pertaining to construction regulations must be strictly adhered to. The protection of life which includes limbs is one of the five objectives of the *Maqasid al-Shariah*.

Establishing of Institutions for Administering, Implementing and Enforcing of Regulations

Islam promotes the system of Caliphate, established based on Shariah as the best system of government to prevent oppression. Because it is based on the rule of law and deprives human beings of arbitrary authority over other human beings, the caliphate system was considered superior to any other (Khaled Abou El Fadl, 2003). In espousing the rule of law and limited government, classical Muslim scholars embraced core elements of modern democratic practice. Limited government and the rule of law are two important aspects of government (Khaled Abou El Fadl, 2003). Democracy's moral power lies in the idea that the citizens of a nation are the sovereign, and in modern representative democracies people can elect their representatives, not merely leaders. In a democracy, the people's fundamental rights must be given protection to ensure their wellbeing and interests of the individual members are guaranteed. Government must establish institutions for administering, implementing and enforcing of law to provide a balanced system.

Islam also recognizes the establishment of institutions and delegation of powers in implementing and enforcing law. This can be seen in the practice of the Prophet (pbuh) sending diplomats in representing him. Any act which is not against the Shariah and that might bring good to the people is not only allowed but encouraged in Islam. A legal maxim *'al-'umur bi maqasidiha* which means "matters are determined according to intention" implies that delegation based on the right intention is allowed in Islam. This legal maxim is formed based on a *hadith* narrated by Bukhari and Muslim, where they said that Umar b. al-Khattab narrated that the Prophet (pbuh) said:

> "Deeds are [a result] only of the intentions [of the actor], and an individual is [rewarded] only according to that which he intends. Therefore, whosoever has emigrated for the sake of Allah and His messenger, then his emigration was for Allah and His messenger. Whosoever emigrated for the sake of worldly gain, or a woman [whom he desires] to marry, then his emigration is for the sake of that which [moved him] to emigrate." (Imam An-Nawawi Forty Hadith).

The persons delegated with the power are also expected to be accountable for their acts and consequences arising from wrongful exercise of powers. The concept of "accountability" is emphasized in Islam. This is because in Islam, the final or primary accountability rests with Allah (*s.w.t*). Enforcement of law and conflict resolution is given utmost importance.

Regulating Professionals Involved in the Construction and Maintenance Sector

The professionals involved in the construction and maintenance of buildings must be provided with guidelines on the conduct of their work that is expected to be carried out in accordance with the required professional standards. Muslims' actions are required to tie in with their intention (*niyyah*). However, the intention must be right to arrive at the true goal. Prophet Muhammad (pbuh) said: "a man's action is according to his intention" (An-Nawawi Forty Hadith: Hadith No.1). Any person involved in planning, construction and maintenance must ensure that his activities are organized strategically to comply with principles of Shariah to achieve pleasure of God (*mardatillah*). A planner must not plan with intention solely to derive profit at the expense of the community and the environment, as this can cause lack of safety of the people and degrade environmental protection.

Specifically, project management in construction encompasses a set of objectives which may be accomplished by implementing a series of operations subject to resource constraints. There are potential conflicts between the stated objectives with regard to scope, cost, time and quality, and the constraints imposed on human, material and financial resources. These conflicts should be resolved at the onset of a project by making the necessary tradeoffs or creating new alternatives. Subsequently, the functions of project management for construction generally include the following:

- Specification of project objectives and plans including delineation of scope, budgeting, scheduling, setting performance requirements and selecting project participants.
- Maximization of efficient resource utilization through procurement of labor, materials and equipment according to the prescribed schedule and plan.
- Implementation of various operations through proper coordination and control of planning, design, estimating, contracting and construction in the entire process.

- Development of effective communications and mechanisms for resolving conflicts among the various participants.

Relying on Shariah principles, the project management personnel is required to observe the ethical values to ensure they conduct themselves in an ethical manner within the permitted limits of Shariah. Professionals are required to observe the three party relationships imposed by Shariah in designing and executing construction plans.

Relationship between Man and the Creator

In the context of observing the relationship between man and his Creator, man is reminded to observe his relationship with his Creator in all aspects of his life. This relationship must be made the focal point of every aspect of man's life in order for man to ensure he protects and maintains the other two aspects. Land planning and development must be organized in a way that it can bring man closer to his Creator. Man must ensure all land development activities are in conformity with the natural laws and not cause any harm to the environment and natural disasters in order to earn the blessings of his Creator.

Relationship between Man and the Environment

Muslims believe that environmental protection is embodied within the central concept of Islam that is *tawhid* (oneness of god), *khilafa* (vicegerency and trusteeship role of man) and *akhirah* (accountability to god in the hereafter) (Hope and Young, 2017). Majority Muslims will undoubtedly take this as a revelation to know how emphatic the Quran is about protecting the environment. The Islamic approach to the environment is holistic. Everything in creation is linked to everything else; whatever affects one thing ultimately affects everything. Man has been distilled from the essence of nature and so is inextricably bound to it (Hope and Young, 2017). The earth's resources such as: land, water, air, minerals, forests are available for our use, but these gifts come from God with certain ethical restraints imposed on the way man uses them (Ghoneim, 2009). We may use them to meet our needs, but only in a way that does not upset ecological balance and that does not compromise the ability of future generations to meet their needs. Islam views the environment as a source of life not only for human beings, but for all living organisms. It is based upon protection, and it encourages revival, construction and development. Allah in Quran 11: 61 proclaimed that:

> "...It was He Who made you from the earth made you thrive upon it...".
> This verse can be interpreted to mean that the Shariah grants concessions
> in terms of taxes to those who take the initiative of reviving deadlands,
> outbacks, and wastelands (*ihya al-mawat*).

Relationship between Man and Man

The relationship between man and man is required to be safe-guarded. Man cannot harm the other and is required to live in harmony with each other. This is the basis of the creation of the society (*ummah*). The Shariah promotes interdependence between man and expects man to respect the privacy of other individuals. In respecting the privacy of his neighbor, man is required to ensure that any development that he proposes will not be violating the neighbor's privacy. Even the minaret of a mosque is not allowed to encroach and violate the privacy of any adjoining land. The rights of a landowner to develop and construct on his land must be exercised cautiously so as not to cause any adverse impact on any existing land use in the neighboring land. Right of way either public or private is very important as it provides access to public terminal. Muslims are required to ensure in the course of developing their land that they allocate at least minimum width of land to allow right of way and not to cause obstruction. Man's obligation towards his neighbor is prescribed in Quran 4: 36 and *Sunnah* (Sahih Bukhari Vol. 8, Hadith No. 6110). Man is required to do good to fellow mankind irrespective of whether he is close or far from him. Shariah requires man to treat his neighbor with equality and justice and not to transgress his rights.

Prophet Muhammad (pbuh) said: *la darar wala dirar fi al-Islam* (no injury should be imposed nor an injury to be inflicted as a penalty for another injury) (Sunan Ibn Majah, Hadith No. 2340). This indicates that no one may inflict harm on himself and another. Significantly, the harm, according to this maxim, is not only limited to causing harm to human beings but the same is applied to the environment, since harm to it contributes to the spread of, and increase in, diseases and thereby threatens everyone's right to a healthy life.

Good Governance and *Hisbah*

In Islam generally, good governance is premised on justice, equity and rule of law. Policy reforms introduced by the government must be in line with *siyasah al-shar'iyyah* (public policy) which will promote sustainable development in the society. Therefore, in the administration of sustainable development institutions in Islam good governance is very important. The mechanisms of good governance in Islamic law include mutual consultation (*shura*), public policy, enjoining good and forbidding evil (*hisbah*), sincere advise (*nasihah*), freedom to criticize (*hurriyyah al-mu'aradah*), freedom to express an opinion (*hurriyyah al-ra'y*) and public interest (*maslahah ammah*) (Mohammad Hashim, 1998). In most cases, the Head of State delegates his powers to his deputies in the execution of governmental policies. However, before any policy is executed, there must be mutual consultation, and high regards must be given to *siyasah al-shar'iyyah* (Mohammad Hashim, 1996). In relation to sustainable development, there are legal texts in the Quran and *Sunnah* which prohibit wastefulness and profligacy in all issues relating to human and material resources that have been provided for man. All the resources with which Allah has endowed mankind are meant to be used, developed and sustained to avoid intergenerational penury among mankind.

The principle of *amr bi al-ma'ruf wa nahy 'an al-munkar* (enjoining people to do good and preventing them from doing bad deeds) is attached to the Islamic concept of good governance which is founded in the institution of *muhtasib* (ombudsman).

While describing the mandate of the governing authority and importance of *hisbah*, Iyad Abumoghli (n.d) rightly observed that the primary duty of the ruler and his assistants, whether they are administrative, municipal, or judicial authorities, is to secure the common welfare and to avert and eliminate injuries to the society as a whole. This includes protection and conservation of the environment and natural resources. Historically, many of the responsibilities of environmental protection and conservation have come under the jurisdiction of the office of the *hisbah*, a governmental agency that was charged specifically with the establishment of good and eradication of harms. The *muhtasib*, who headed this office, is required to be a jurist thoroughly familiar with the rulings of Islamic law that pertained to his position (Ibn Taimiyyah, 1992).

Mohammad Hashim (1996) further highlighted the duties of the *muhtasib* as being responsible for the inspection of markets, roads, buildings, watercourses, reserves (*hima*) and so forth, all areas making up the built environment.

The protection of proprietary rights of individuals and the governance system in Islam is briefly summarized by Zamir and Abbas (2005) as, "the design of governance system in Islam can be best understood in the light of principles governing the rights of individuals, society and the state, the laws governing propriety ownership, and the framework of contracts. Islam's recognition and protection of rights is not limited to human beings only but encompasses all forms of life as well as the environment." Each element of Allah's (*s.w.t*) creation has been endowed with certain rights and each is obligated to respect and honor the rights of others. These rights are bundled with the responsibilities for which humans are held accountable.

The system of governance in Islam gives all its citizens rights and corresponding duties. The right to acquire property is part of the fundamental rights granted to citizens in the State. However, the government ensures that those properties privately owned are used judiciously for the benefit of one's self or the Muslim *ummah* at large. If the economic viability of such property can be made, then the government must lay down policies that will ensure sustainable development. Furthermore, the government must maintain social justice in the society by assisting the underprivileged and marginalized have-nots in developing their privately-acquired properties.

Muslims must establish their leadership role in furthering the sustainable development institutions for the benefit of mankind. This goal can be pursued through good governance in all sectors of life and full utilization of resources without being wasteful. The idea of moderation and balanced community, as emphasized in the Quran, has been amplified by Zubair (2006) where he observed that moderation and balance in worldly pursuits that the verses of the Quran repeatedly emphasize are intended to support the basic Islamic concept of sustainable development. The achievement of the *maqasid* (goals) calls for dynamic interaction between socioeconomic processes and environmental priorities.

It is clear that good governance and sustainable development are two inextricable concepts that are closely linked. It is therefore imperative for Muslims to unravel the treasure of the ideals of Islam in relation to sustainable development for proper administration of sustainable development institutions in the modern world.

Regulatory Institutions and Protection of Social Security

A governing authority must ensure every citizen is given the right to education, health, healthy environment and good standard of living which requires promotion of efficient construction practices. Social security is a fundamental part of good governance which can be promoted through proper channelling of State resources in the development of slums and land inhabited by the poor. It is the duty of the State to introduce policies that will assist in poverty alleviation by providing affordable living.

Iyad Abumoghli (n.d) confirms this by saying that the Islamic way of poverty alleviation focuses on developing human resource and providing relevant job opportunity. The institutions identified for financial assistance to the poor are assisted (*kifalah*) by the nearest kin; the neighbors under neighborhood rights; others in the form of mandatory charity like obligatory contribution (*zakat*); and through temporary and permanent endowments. Moreover, a State is bound to provide sustenance to its citizens irrespective of their religion. The State meets this responsibility by collection of *zakat*, other emergent charities and raising taxes. The extent of such relief to the poor under Islam cannot be disputed.

As earlier discussed, institutions of *zakat, bait al-mal* and *wakaf* are important mechanisms for alleviation of poverty when properly managed. According to Abd al-Rahman (2005) "the motivation for poverty alleviation and human development in Islam is linked with Shariah and *aqīdah*. Shariah imposes a fiscal duty on production as well as on idle liquid wealth, because *zakāh* and *aqīdah* motivate believers to pay charity (*sadaqah*) to those who are not able to meet the basic necessities of life." The aim of these fiscal policies is not only to meet the basic needs of the poor but to remove them from the poverty bracket and transform them to wealth-creating entities through sustainable developmental policies.

The right to healthy living and clean environment is an important aspect of Islam. The fundamental basis of the five canonical prayers in Islam is purity of body and mind. One cannot proceed to observe a prescribed prayer (*salat*) without undergoing a particular form of purification in the form of ablution. This is observed at least five times a day. This speaks volumes about the high premium Islam places on purity and the need to live in a healthy environment. This is closely related to right to good standard of living and good health facilities. The governing authority must ensure enough food supply and adequate water supply for the benefit of all. Epidemics and diseases must be curtailed and health facilities must be accessible to all. The use of chemicals and other elements that may stand as potential risks to the masses must be controlled, while sewage and domestic waste must be well treated and, if possible, recycled in line with true Islamic ideals which encourage sustainable development and condemn profligate depletion of natural resources (Abu-Lughod and Janet, 1993; Ahmed Farid, 1986).

Stakeholders Participation in Promoting Good Construction Practices

Participation in decision-making or *shura* is taken here to mean the evolving decision-making process at all levels of the Islamic society.

The practice of *shura* has its meaning in the Quran: *Wa amruhom shura baynahum*. The Quran advises man to maintain the balance as the world was created in balance:

> "And (remember) We did not create the heavens and the earth and all that lies in between in mere idle play. Had it been Our will to find a pastime, We could have found one in Our presence, if We were to do it at all." (Quran 21: 16-17).

> "He has created man; He has taught him articulate speech. The sun and the moon pursue their appointed courses according to certain order and computation; And the herbs and the trees — both prostate themselves submitting to His Ordered course. And the heaven He has raised high, and He has set up the rules and order for the Balance of Justice, In order that you may not transgress (the) Balance of Justice; And give just weight and fall not short in balance." (Quran 55: 3-9).

The communities' role in promoting sustainable practices can be inculcated with providing opportunities to contribute in the decision-making process and question poor practices to promote efficiency within the construction industry.

Promote *Itqan* (Perfection) and *Ihsan* (Compassion)

The quality management system pioneered by the Prophet (pbuh) and his companions is universal in nature, as it includes two dimensions namely, material (*lahiriyah*) and spiritual (*ruhiyyah*). The system emphasizes culture of perfection (*itqan*) and the values of compassion in the performance of an action whether in congregational worship or individual worship such as prayer, fasting, charity and others. Based on these traditions, we may relate to this culture of perfection as a culture of excellence, earnest and steadfast (commitment), generated through diligent work environment that is comfortable and conducive. Thus, workers are able to produce quality products or services and achieve customer satisfaction. Hence, in this context, continuous improvement can be equated to the concept of *islah* (to perform or improve), which is very important in encouraging innovation and pursuit of excellence. According to Azman (2003), *itqan* is the culture that often refers to the implementation and completion of the work carried out diligently and earnestly. Sohaimi and Gunawan (2007) define the concept of *itqan* as a commitment to achieve perfection. A manager who practices *itqan* culture will feel confident that every task entrusted to him is committed with accountability because of the belief he will be rewarded in this world and the hereafter. Therefore, when a person makes the culture of perfection an important element in executing his assignments, his work will be of high quality. In order to make the culture of perfection commendable practices that should be adopted by the managers or employees of an organization, they must first understand their roles in fulfilling the individual rights of Muslims. The rights of the employer, workers, colleagues, subordinates and others in the organization must be fulfilled as far as possible. The concept of perfection (*itqan*) should be based on (Ab. Mumin and Fadillah, 2006):

(1) The establishment of the relationship between man and the Creator (Tawhid) through demonstration of outstanding work and constant effort in increasing their knowledge and skills to keep up with the latest development and change.

(2) Each employee, either the management or the staff (executor) must demonstrate a positive attitude or become a good role model to be emulated by others. This is essential as the leader must establish virtues and values to be emulated by the employees. A good leader does not depend merely on power or influence, but must demonstrate good virtues and values, good moral conduct and work ethics and superior appearance.

(3) The constant practice of change or transformation in the performance of daily work like creating new things or innovation to simplify the execution to be more effective to address the changes in the organization due to changing needs of technology, globalization and competition from others.

Managers should be skilled to manage any changes in the organization and be able to deal with conflicts effectively through the guidance of the Quran and *hadith* (Khaliq Ahmad and Fontaine, 2011; Khaliq Ahmad, 2006).

Education and Raising Awareness

The teachings of Islam have an ethical notion that guides Muslims to care about the environment; knowledge that helps them perfect their duties:

> "And He taught Adam the names of all things and their uses; then He showed them to the angels, and said: 'Tell Me the names of these (things), if what you say is true.'" (Quran 2: 31).

This verse describes how and why humankind was given the ability to know the names of creation. It is an important symbol of knowledge given only to the human race from among all the other creatures including angels. Therefore, using religious education to convey the messages of importance of conducting efficient construction activities is an excellent tool.

Conclusion

Muslims are faced with a task of developing the *ummah*'s wellbeing. They are often made to believe that Islamic law is very limited and it is not easy to derive law ordinarily. Hence, many Muslim nations prefer to adopt the proven models of Western law. However, they are beginning to realize that most of these models failed to meet the holistic needs of Islam, and had persistently proven to be unsustainable. The Shariah offers solutions for man to live in harmony with nature. Man must be aware of the *barakah* he will earn by promoting good deeds and preventing harm to the society and his responses and priorities should be based on the moral authority of being of service to humanity. The Quran in 3: 104 provides:

> "Let there be among you a community of people who shall call to all that is good (to spread Islam), enjoin what is right, and forbid evil. Such men shall surely triumph."

Shariah principles can provide very useful values that can help improve the present predicament faced by the construction industry.

List of Cases

Beximco Pharmaceuticals Ltd & Others v. Shamil Bank of Bahrain EC [2004] EWCA Civ 19.

Commissioner of Police v. Acharya Jagadishwaranada Avadhuta [2004] 2 LRI 39.

Hajjah Halimatussadiah bte Hj. Kamaruddin v. Public Services Commission Malaysia & Anor [1994) 3 MLJ 61.

Halpern v Halpern & Anor [2007] EWCA Civ 291.

Javed v State of Haryana (AIR) [2003] SC 3057.

Meor Atiqulrahman Ishak & Ors v Fatimah Sihi & Ors. [2006] 4 CLJ 1.

Musawi v. Re International (UK) and Others [2007] EWHC 2981 (Ch).

Sardar Syedna Taher Saifuddin Saheb v. State of Bombay AIR [1962] SC 853.

The Commissioner, Hindu Religious Endowments, Madras, v. Sri Lakshmindra Thirtha Swamiar AIR [1954] SC 282.

References

Abdulac, S (1984). Large-scale development in the history of Muslim urbanism. In *Continuity and Change: Design Strategies for Large-Scale Urban Development*, MB Sevcenko (eds.), pp. 2–11. Cambridge, Massachusetts: The Aga Khan Program for Islamic Architecture.

Ab. Mumn, AG and M Fadillah (2006). *Dimensi Pengurusan Islam: Mengurus Kerja dan Mengurus Modal Insan.* Kuala Lumpur: Universiti Malaya Publication.

Abd al-Rahman, Y (2005). Sustainable development: An evaluation of conventional and Islamic perspectives. In *Islamic Perspectives on Sustainable Development*, Munawar Iqbal (ed.). New York: Palgrave Macmillan.

Abu-Lughod and L Janet (1993). The Islamic city: Historic myth, Islamic essence, and contemporary relevance. In *Urban Development in the Muslim World*, A Hooshang and SES Salah (eds.), pp. 11–36. New Brunswick, NJ: Center for Urban Policy Research.

Ahmed Farid Mustapha (1986). Islamic values in contemporary urbanism. A paper presented at First Australian International Islamic Conference, organised by the Islamic Society of Melbourne, Eastern Region (ISOMER).

Amini Amir, A (1997). Konsep Seni dalam Islam. In *Tamaddun Islam*, Ahmad Fauzi Hj. Morad and Ahmad Tarmizi Talib (eds.). Serdang, Penerbit Universiti Putra Malaysia.

Azman, CO (2003). *Pengurusan Di Malaysia dari Perspektif Islam.* 3rd Ed. Kuala Lumpur: Dewan Bahasa dan Pustaka.

Baster, N (1972). *Measuring Development*. Frank Cass, London.

Beg, MAJ (1983). *Two Lectures on Islamic Civilisation*, pp. 28–29. Kuala Lumpur: University of Malaya Press.

Besim Selim Hakim (1986). *Arabic–Islamic Cities, Building and Planning Principles*. London: KPI Limited.

de Montequin, FA (1980). The essence of urban existence in the world of Islam. Paper presented at the Symposium on Islamic Architecture and Urbanism, King Faizal University Saudi Arabia 1980.

Faludi, A (1973). *A Reader in Planning Theory*. New York: Pergamon.

Fazlun, MK (2002). Sustainable development and environmental collapse: An Islamic perspective. Paper presented at the World Summit on Sustainable Development parallel event Muslim Convention on Sustainable Development National Awqaf Foundation of South Africa, September 1, 2002.

Ghoneim, KS (2009). *The Quran and the Environment*. Faculty of Science, Al-Azhar University, Cairo. http://www.islamonline.net/english/Science/2000/4/article/shtml [Accessed 15 November 15, 2017].

Hamid, ME (1969). Islamic law of contract or contracts?, *Journal of Islamic Comparative Law*, 3, pp. 1–11.

Hope, M and J Young (2017). *Islam and Ecology*. http://www.crosscurrents.org/islamecology.htm [Accessed July 20, 2017].

Ibn Taimiyyah (1992). *Public Duties in Islam — The Institution of the Hisbah*. Translated from the Arabic by Muhtar Holland, London: Islamic Foundation.

Ibrahim, A (1988). Some evolutionary and cosmological aspects to erly Islamic town planning. In *Theories and Principles of Design in the Architecture of Islamic Societies*, MB Sevcenko (ed.), pp. 57–72. Cambridge, Massachusetts: Aga Khan Program for Islamic Architecture.

Imam Nawawi. *Forty Hadith*. http://40hadithnawawi.com/ [Accessed August 1, 2017].

Ismawi, Z (1996). Vision of an Islamic city. Unpublished conference paper presented at the Conference on Shaping the Vision of a City, Ipoh: September 16–17, 1996.

Iyad Abumoghli (n.d). *Sustainable Development in Islamic Law*. http://waterwiki.net/images/8/85/Sustainable_Development_in_Islamic_Law_-_Iyad_Abumoghli.doc [Accessed July 28, 2017].

Jasser, A (2008). *Maqasid Syariah Guide*. United States: International Institute of Islamic Thought (IIIT).

Khaled Abou El Fadl (2003). *Islam and the Challenge of Democracy*. Originally published in the April/May 2003 issue of Boston Review. http://bostonreview.net/BR28.2/abou.html [Accessed January 10, 2012].

Khaliq Ahmad (2006). *Management from Islamic Perspective*. Selangor: Research Centre, International Islamic University of Malaysia.

Khaliq Ahmad and R Fontaine (2011). *Management from Islamic Perspective*. Selangor: Pearson Malaysia Sdn. Bhd.

Kurshed, A (1980). *Economic Development in an Islamic Framework*. The Islamic Foundation.

Mawil Izzi, D (2004). *Islamic Law: From Historical Foundations to Contemporary Practice*. Edinburgh: Edinburgh University Press.

Mejelle (2000). *A Complete Code of Islamic Civil Law*. Translation of Majallah el-Ahkam-i-Adliya. Kuala Lumpur: A.S. Noordeen.

Mohammad Hashim, K (1996). Methodological issues in Islamic jurisprudence. *Arab Law Quarterly* (London), 1, 3–34.

Mohammad Hashim, K (1998). *Freedom of Expression in Islam*. Kuala Lumpur: Ilmiah Publishers Sdn. Bhd.

Mohammad Hashim, K (2009). *Maqasid al Shariah: The Objectives of Islamic Law*. Kuala Lumpur: International Institute of Islamic Thought (IIIT).

Muhammad H (1999). *Maqasid al-Shariah: The Objectives of Islamic Law*. Pakistan: International Islamic University Islamabad.

Moustapha, AF (1986). Islamic values in contemporary urbanism. Paper presented at the First Australian International Islamic Conference organised by the Islamic Society of Melbourne, Eastern Region (ISOMER). (Unpublished).

Portney, KE (2003). *Taking Sustainable Cities Seriously*. Massachusetts: Massachusetts Institute of Technology.

Rabah, S (2003). *Introduction to the Islamic City*. Foundation for Science Technology and Civilisation (FSTC) Limited. http//www.muslimheritage.org.

Rispler-Chaim, V (2007). *Disability in Islamic Law*. The Netherlands: Springer.

Sayyed Hossein, N (1968). *Man and Nature: The Spiritual Crisis in Modern Man*. London: George Allen and Unwin.

Schacht, T (1950). *The Origins of Mahammadan Jurisprudence*. Oxford: Clarendon Press.

Sohaimi HMS and CAA Gunawan (2007). Kualiti dari perspektif Islam: Satu pengamatan dalam perkhidmatan awam Malaysia. Main presentation paper for Quality Day Celebration, the Commision of Malaysia Civil Service on December 7, 2007.

The Translation of the Meanings of Sahih Al-Bukhari Arabic–English Volume 8 (1997). Translated by Muhammad Muhsin Khan. Riyadh, Saudi Arabia: DARUSSALAM Publishers and Distributors.

Tarek Badawy (2012). The general principles of Islamic law as the law governing investment disputes in the Middle East. *Journal of International Arbitration*, (29)3, 255–267.

Yusuf al-Qaradhawy (2009). *Agama Dan Politik: Wawasan Ideal dan Menyanggah Kekeliruan Pemikiran Sekular-Libral*. Kuala Lumpur: Alam Raya Enterprise.

Zamir I and M Abbas (2005). The stakeholders model of governance in an Islamic economic system. In *Islamic Perspectives on Sustainable Development*, Munawar Iqbal (ed.). New York: Palgrave Macmillan.

Zubair, H (2006). Sustainable development from an Islamic perspective: Meaning, implications and policy concerns. *Journal of King Abdulaziz University: Islamic Economics*, (19)1, (2006 A.D/1427 A.H), 9.

Chapter 4

Shariah-Compliant Construction Marketing: Development of a New Theory[1]

KHAIRUDDIN Abdul Rashid and
Christopher Nigel PREECE

Introduction

In Islam, the way people lead their lives including in conducting personal and business transactions falls under the precepts of Shariah. In the context of business transactions the broad principles within the Shariah that is applicable is *al-muamalat*.

Al-muamalat (singular *muamalah*) or transactions between people emphasizes the need for business transactions to apply the concepts of justice, moral obligation, accountability and equality; these aspects are in line with the Islamic belief (*al-iman*), practices (*al-amal*) and value system.

According to Khairuddin (2007, 2008) *muamalat* in the areas of banking, insurance and finance and capital markets has shown phenomenal success. However, he argued that *muamalat* in the

---

[1] This chapter has its origin from Preece and Khairuddin (2009). Shariah Compliant Construction Marketing — Development of a New Theory. In Khairuddin *et al.* (2009). Ed. *Collaborative Efforts in International Construction Management*. Proceedings, IIUM and Kyoto University, October 21, 2009, Kuala Lumpur.

area of construction contracts — a contract between the client and contractor; the client and consultants; contractor and sub-contractors or suppliers, etc. — is still in its infancy. In addition, he observed that the enthusiasm shown by scholars and professionals, muslims and non-muslims alike, to translate relevant Islamic theories and principles contained in the Quran and the *Sunnah* of the Prophet Muhammad (pbuh) into working models suggests this phenomenon is set to change.

Similarly, it is contended herein that *muamalat* in the area of construction marketing — professional services and products — is also in its infancy. This chapter discusses the concept of Shariah-compliant construction marketing, identifies key elements thereto and explains why the concept is considered significant and worthy for further consideration.

What Is Marketing?

Marketing is defined by the American Marketing Association (2017) as the activity, set of institutions, and processes for creating, communicating, delivering and exchanging offerings that have value for customers, clients, partners and society at large. The term, developed from the original meaning, refers literally to going to a market, as in shopping, or going to a market to buy or sell goods or services.

The Chartered Institute of Marketing (2017) in the UK defines marketing as "The management process responsible for identifying, anticipating and satisfying customer requirements profitably."

Marketing practice has tended to be seen as a creative industry in the past, which included advertizing, distribution and selling. However, as marketing makes extensive use of social sciences, psychology, sociology, mathematics, economics, anthropology and neuroscience the profession is now widely recognized as a science. The overall process starts with marketing research and goes through market segmentation, business planning and execution and ending with pre- and post-sales promotional activities. It is also related to many of the creative arts.

In recent years, a new approach to ethical marketing has been developing. Ethical marketing may be defined as an honest and factual representation of a product, delivered in a framework of cultural and social values for the consumer (Laczniak *et al.*, 1978; Murphy *et al.*, 2005). Kotler (2003) maintains that this new "societal marketing concept holds that the organization's task is to determine the needs, wants and interests of target markets, and to deliver the desired satisfactions more effectively and efficiently than competitors in a way that preserves or enhances the consumer's and the society's wellbeing."

The concern with ethical issues, such as child labor, working conditions, relationships with third world countries and environmental problems, has changed attitudes towards a more socially responsible way of thinking (Fan, 2005). This has influenced companies and their response is to market their products in a more socially responsible way. The increasing trend of fair trade is an example of the impact of ethical marketing. The philosophy of marketing is not lost, but rather hopes to win customer loyalty by reinforcing the positive values of the brand (Fan, 2005).

Ethical marketing should not be confused with government regulations brought into force to improve consumer welfare, such as reduce carbon dioxide emissions to improve the quality of the air. Enlightened ethical marketing is at work when the company and marketer recognize further improvements for humankind unrelated to those enforced by the government. By way of example, the Coop Group in the UK refuses to invest money in tobacco, fur and in any countries with oppressive regimes (Ginsburg, 2006).

What Is Construction Marketing?

Marketing in the construction industry is still at an embryonic state of development. The industry became aware of the need for more professional marketing and sales efforts in line with other industries during the 1980s and the recession of the 1990s accelerated further the process as organizations were faced with new challenges (Smyth, 2001). The construction sector has seen significant changes with increasing privatization, greater client demands for quality, attempts to change the adversarial culture through partnering and internationalization. The private sector has expanded and domestic and foreign competition has increased.

Marketing in the construction sector is largely business-to-business rather than business-to-consumer (Pettinger, 1998). This kind of market is typified by fewer and larger buyers, many participants in the buying process, professional and perhaps more rational buyers, and a closer relationship between supplier and buyer. The suppliers of the industrial product or service often influence the management practices of the client (Filiatrault and Lapierre, 1997).

Smyth (2001), Collard and Preece (2000), and Macnamara (2003) agree that relationship marketing, which was developed for business-to-business markets, primarily in the service sector (Gronroos, 1996), is an appropriate concept to apply to firms in construction. The relationship marketing concept is defined as the process of identifying and establishing, maintaining, enhancing and when necessary terminating relationships with customers and other stakeholders (Gronroos, 1996). Relationship marketing is a concept for developing long-term and sustained contact with clients or customers so that their needs can be targeted and satisfied in return for client loyalty. The benefit for the supplier is repeat business or high levels of orders in referral markets (Smyth, 2001).

What Is Marketing Mix and Its Application to Construction, i.e. The 5 Ps of Product, Price, Promotion, Place and People?

The traditional marketing mix of key decision areas was invented for the consumer manufacturing sectors and included four elements (Culliton, 1948; Borden, 1964), the so-called 4 Ps of marketing: Product, Price, Promotion and Place. With the growth of service sectors and industrial marketing these were seen to be inadequate and a fifth P representing those decisions concerning the service providers, the people involved in the transaction, was included (Bitner and Booms, 1981; Barlon, 2006).

A number of writers and researchers in construction marketing have attempted to apply and develop marketing models from other sectors (Smyth, 2001; Macnamara, 2003). Preece (2001) applied the 5 Ps model to construction service businesses and identified a number of challenges to decision-making as follows:

- **Product** — contractual services are difficult to standardize. They are heavily reliant on the people within the business. The construction marketer needs to identify the features of the services and how these provide clear benefits to the client. These are essentially concerned with price, time and quality.
- **Price** — services are difficult to price before the service has been completed.
- **Promotion** — unlike tangible goods such as cars or televisions, contractual services cannot be displayed. The features and benefits of the service have to be demonstrated through promotional activity. Promotion makes the intangible, tangible, by creating an image for the customer of the service he should receive.
- **Place** — this is where the service is performed for the client, e.g. at the client's office or the project site.
- **People** — the service providers. Many managers and staff throughout a construction business will have contact with the clients or customers on site, in the office or over the telephone. Many people are involved throughout the business network: subcontractors, suppliers, consultants, local communities, a variety of stakeholders. The success of the project and the quality of the overall service depends on how these individuals, groups and organizations interact.

Traditionally the construction industry has had an adversarial professional and business culture which has frustrated successful implementation of the marketing concept. The application of marketing planning and the 5 Ps model can still be seen to be at an embryonic stage, particularly in the contracting and consulting sectors. Yisa *et al.* (1995) and Preece and Barnard (1999) established that it was mainly the larger companies that have a formalized marketing function. Few contractors or consultants employ individuals with marketing qualifications. A serious problem would seem to be the lack of planning and long-term focus. For example, Morgan and Morgan (1991) found that only a quarter of consultancies carried out market research on a regular basis.

The Shariah Perspective of Construction Marketing

Given that Islam views conducting businesses that comply with the Shariah as a general worship and such activities will receive reward from Allah the Almighty, the practice of marketing should also be in line with the Shariah. Broadly therefore, the practice of construction marketing should be in line with the principles of *al-muamalat*.

In the current work Shariah-compliant marketing refers to the practice of marketing that embraces the Islamic doctrines and reiterates the Islamic belief, practice and value system, etc. It may be briefly illustrated by examining the conventional 5 Ps model on marketing: product, promotion, price, place and people (see Table 4.1).

Why Shariah-Compliant Construction Marketing?

The idea to develop Shariah-compliant construction marketing stems from the many critiques of the current practice in construction marketing. They include:

(1) The product or service delivered did not often live up to the promises made by firms or satisfy the expectations of clients from a quality perspective;
(2) In terms of price, it was considered there was often a lot of price uncertainty and value for money was often not achieved;
(3) High levels of corruption, bribery and unethical behavior which affect marketing activities across the industry;
(4) In terms of the people element, respondents noted that firms were known to use unqualified people to certify elements of projects which had sometimes led to disasters; and
(5) Lack of attention to safety on construction projects.

Shariah compliance in construction marketing is a good thing, as such a practice could improve the workings of the construction

Table 4.1. Developing the Elements of Shariah-Compliant Construction Marketing.

5 Ps	Principles of Shariah compliance
Product	The product must be clearly defined, illustrated, quantified and specified so as to avoid ambiguity or misrepresentation. . . the time of completion must be specified. . . the materials to be used in the works should be supplied by the contractor. The product, upon its completion should not be associated with elements contrary to the Shariah such as trading in alcohol, gambling, prostitution, etc.
Price	The price to be paid to the parties concerned must have value, e.g. the consideration being defined in the form of the contract sum and in authorized currency. The price to be paid by the client to the consultants or constructors should not contain elements of *gharar, maysir, riba, al-ihtikar, iktinaz* (hoarding, black marketeering) and *talaqqi al-rukban* (middle person; not *al-ijara'* or one who serves others) leading to deception or inflated prices of goods or commodities are forbidden.
Promotion	In promoting a product or service, all known faults of the product or service must be made known to the clients or customers thus avoiding *gharar* and *maysir*. Consultants and the contractors must avoid making false assertions; false testimonies; deception; inducing the client to make payments through cunningness and craftiness; the giving of sexual favors or other forms of immoral activities including bribery or instilling emotional fear or similar acts.
Place	The promoter or initiator must possess ownership or have the authority to develop the land or building on which the completed facility will occupy, and has obtained all necessary approvals.
People	Parties in a contract or covenant must be "qualified," and must fulfil their respective obligations. Allah commands "O believers! Be true to your obligations. . ." (Quran 5: 1). In conducting construction businesses the concepts of truthfulness, ethics, integrity and accountability must be upheld at all times.

industry as a whole. The benefits of Shariah-compliant construction marketing include:

(1) Greater transparency and fairness to all parties involved in projects;

(2) Minimization of disputes and projects can be executed according to specifications and to meet clients' requirements;

(3) Increase productivity and quality of the product;

(4) People working in the industry would benefit from practicing good ethics;

(5) Creating a fairer and less corrupt culture; and

(6) If countries such as Malaysia developed Shariah-compliant construction marketing, then it could encourage further interest and investment from other Islamic countries.

Conclusion

This chapter examined the concepts of construction marketing as it is being practiced in the conventional way and Shariah-compliant construction marketing. In the latter, the conventional marketing mix for service sectors, i.e. the 5 Ps were infused with elements of *al-muamalat* so that the new 5 Ps would become consistent with the requirements of the Shariah.

It should be borne in mind that this chapter is only the start of the development of this new theory. The outcome of the limited research so far, in terms of the rather negative critiques on the current practice of construction marketing, and of the positive interest shown in the concept of Shariah compliance, provide impetus for the authors to continue with this research.

References

American Marketing Association (2017). Definition of Marketing. https://www.ama.org/AboutAMA/Pages/Definition-of-Marketing.aspx [Accessed October 26, 2017].

Barlon, K (2006). The concept of the marketing mix. Presentation on marketing management at Turku University, Finland, September 2006. The same article can also be found in: Schwartz, G. (ed.) (1965). *Science in Marketing*, pp. 386–397. New York: John Wiley; and also in: Enis, B and Cox, K (1991). *Marketing Classics, A Selection of Influential Articles*, pp. 361–369. Boston: Allyn and Brown.

Bitner, J and B Booms (1981). Marketing strategies and organizational structures for service firms. In *Marketing*, J Donnelly and W George (eds.). Chicago: American Marketing Association.

Borden, NH (1964). The concept of the marketing mix. *Journal of Advertising Research*, June(4), 2–7. Available in Schwartz G (1965). *Science in Marketing*, pp. 386–397. New York: John Wiley and Sons.

Collard, P and CN Preece (2000). *Guide to Marketing Terminology for Construction Professionals*. United Kingdom: Marketing Works Training and Consultancy.

Culliton, JW (1948). *The Management of Marketing Costs*. Harvard University: Graduate School of Business Administration.

Fan, Y (2005). Ethical branding and corporate reputation. *Corporate Communication*, 10(4), 341.

Filiatrault, P and J Lapierre (1997). Managing business-to-business maketing relationships in consulting engineering firms. *Industrial Marketing Management*, 26, 213–222.

Ginsburg, RS (2006). *Ethical Marketing Skills for Lawyers*. Denver: Continuing Legal Education in Colorado. OCLC 133147723.

Gronroos (1996). Relationship marketing: strategic and tactical implications. *Management Decisions*, 34(3), 236–238.

Khairuddin, AR (2007). Shariah compliant contract: A new paradigm in multinational joint venture for construction works. In *International Joint Ventures: Reaching Strategic Goals*, K Kobayashi *et al.* (eds.), pp. 14–25. Bangkok: Asian Institute of Technology Thailand and Kyoto University Japan.

Khairuddin, AR (2008). Shariah compliant contract for construction works: Setting the agenda for research. In *Proc. Conf. Shariah Compliant Construction Contract*, International Islamic University Malaysia. Kuala Lumpur: April 16, 2008.

Kotler, P (2003). *Marketing Management*, 11th Ed. New Jersey: Prentice Hall Education.

Laczniak, ER, FL Robert and AS William (1978). *Ethical Marketing: Product vs. Process*. Madison: Graduate School of Business, University of Wisconsin-Madison.

Macnamara, P (2003). Marketing of civil engineering consultants in the UK. In *Construction Business Development: Meeting new challenges, seeking opportunity*, CN Preece, K MoodleyP Smith P (eds.), pp. 39–68. Butterworth-Heinemann.

Morgan, RE and NA Morgan (1991). An appraisal of the marketing development in engineering consultancy firms. *Construction Management and Economics*, 9, 355–368.

Murphy, PE, RL Gene, EB Normal and AK Thomas (2005). *Ethical Marketing*. Upper Saddle River, New Jersey: Pearson Prentice Hall.

Pettinger, R (1998). *Construction Marketing — Strategies for Success*. London: Macmillan Press Ltd.

Preece, CN (2001). Marketing and promotional strategies in construction. In *Strategic Management in Construction*, 2nd Ed, D Langford and S Male (eds.), pp. 175–191. Oxford: Blackwell Science.

Preece, CN and L Barnard (1999). *Report on the State of Marketing in UK Engineering Consultancies*. University of Leeds: Construction Management Group, School of Civil Engineering.

Smyth, H (2001). *Marketing and Selling Construction Services*. Oxford: Blackwell Science Limited.

The Chartered Institute of Marketing, UK (2017). *Marketing and the 7Ps: A Brief Summary of Marketing and How It Works.* https://www.cim.co.uk/media/477 2/7ps.pdf [Accessed October 26, 2017].

Yisa, SB, IE Ndekugri and B Ambrose (1995). Marketing function in UK construction contracting and professional firms. *Journal of Management and Engineering,* 11(4), 27–33.

Part 2

Shariah-Compliant Construction Contract: Concept and Application

Chapter 5

Shariah-Compliant Contract: Concept and Application for Construction Works

KHAIRUDDIN Abdul Rashid

Introduction

In Islam, when parties enter into a contract they must ensure that the subject matter, agreement, terms, conditions and the nature of the contract or covenant should be made in accordance to the Shariah; hence the term "Shariah-compliant" (Khairuddin, 2009, p. 103). In other words, Shariah-compliant construction contract is "a construction contract with its subject matter, agreement, terms and conditions that embrace the Islamic belief, practice and value system" (Khairuddin, 2007).

However, in the context of construction works the current practice of applying the conventional-styled contracts remained. Khairuddin (2009, p. 104) contended that *al-muamalat* is still in its infancy insofar as construction procurement — and specifically in contract for construction works — are concerned. Therefore, there is a need for the concept of Shariah-compliant construction contract to be developed and its application promoted. This chapter represents a modest effort towards fulfilling this need.

This chapter is structured into four parts: (i) firstly, it discusses the broad principles of *al-muamalat* and its application in business transactions; (ii) secondly, current contract practice for construction works is examined focusing on the primary contract between the Employer and the Contractor and shows how such a contract varies from the Shariah; (iii) thirdly, key provisions in the current contract agreement for construction works are examined to ascertain their compliance or otherwise with the Shariah; and (iv) finally the fourth part concludes the chapter.

Muamalat and Its Application in Business Transactions

Muamalat refers to the set of rules for business transactions between people. Essentially, business transactions must apply the concepts of justice, moral obligation, accountability and equality in accordance with the Islamic belief (*al-iman*), practices (*al-amal*) and value system (Muhammad Rawwas, 2005, p. 7; Mohd Ma'sum, 2006, pp. 18–19; Khairuddin, 2009, p. 106). The following are some of the verses from the Quran reiterating the importance of these concepts:

> "...and give full measure and weight – in justice..." (Quran 6: 152).

> "And give just weigh and fall not short in the balance." (Quran 55: 9).

> "But if the debtor is in a hard time, grant him a delay until he can pay his debt. But if you remit the debt as alms, it is better for you, if only you know (the generous reward that you will receive)." (Quran 2: 280).

> "O believers! Do not consume (use) your wealth among yourselves illegally (such as means of cheating, gambling and others of illegal nature), but rather trade with it by mutual consent..." (Quran 4: 29).

> "...and fulfill every covenant (to Allah and men). Surely every covenant will be inquired into." (Quran 17: 34).

Muhammad Rawwas (2005, pp. 1–12) listed the following basic principles of *al-muamalat*:

- Business transactions are not forbidden unless otherwise specified by the *nass* (textual evidence found in the Quran and the *Sunnah*) that a certain matter is prohibited;
- The Shariah facilitates (not constrains) people in conducting their daily personal and business lives (*maslahah*);
- *Iktinaz* (hoarding, black marketeering) and *talaqqi al-rukban* (middle person; not *al-ijarah* or one who serves others) leading

to deception or inflated prices of goods or commodities are forbidden;
- Business transactions that involve elements contrary to the Shariah such as *riba* (interest charges), trading in alcohol, gambling, prostitution, etc. are forbidden;
- In the interest of the *ummah*, monopoly (*al-ihtikar*) must be avoided;
- Business transactions should be conducted with patience, tolerance and with the intention to facilitate not to constraint;
- In conducting business, the concepts of truthfulness, ethics, integrity and accountability must be upheld at all times;
- *Gharar* (uncertainty) and *maysir* (gambling) such as market manipulation and harmful speculation are forbidden;
- Parties in a contract or covenant must fulfill their respective obligations. Allah's commands on this matter are very clear thus *"O believers! Be true to your obligations..."* (Quran 5: 1).
- Be diligent and persevere in conducting businesses in such a way that all responsibilities and obligations are discharged in accordance to one's best ability.

Contract in Al-Muamalat

In Arabic, the term contract is referred to as *'aqd*. Basically, it means two or more parties mutually agree to tie, to knot, to conjunct (Mohd Ma'sum, 2006) on a subject matter. The elements of *'aqd*:

(1) *Sighah* — comprising of *ijab* (offer) made by the first party and *qabul* (acceptance) by the second party;
(2) *Aqidan* — persons making the contract must possess the following qualifications: a sound mind, has reached puberty, is free to enter into a contract, and is not a bankrupt; and
(3) *Ma'aqud Alaid* — the subject matter should be in existence at the time of the *al-aqad*, legally owned, beneficial to the contracting parties and have commercial value in accordance to the Shariah; and the consideration that has value (*mutaqawwim*), is certain (*mawjud*), permissible (*halal*) and valid (*sahih*).

A contract between parties should be signed and witnessed. These requirements are stated in the Quran thus:

> "O believers! When you contract a debt from one another for a fixed period, put it (its amount and period of repayment) in writing. And let a scribe write it down between you justly (truthfully), and no scribe should refuse to write as Allah has taught him… and call in two of your men as witnesses…" (Quran 2: 282).

Table 5.1 lists the dominant Islamic banking, insurance and finance and capital markets' contracts or products that are in compliance with the Shariah.

Table 5.1.   Shariah-Compliant[1] Contracts for Business Transactions.

Product	Style of contract	Brief description
Al-Mudharabah	Partnership	A contract of partnership between parties, the borrower (*mudareb*) borrows money from the financier (*rab-al-amal*). The proceeds or otherwise from the venture is to be shared among the parties in accordance with the terms and conditions of the contract.
Al-Bai bithaman Ajil	Sale and purchase	A contract of sale and purchase of goods whereby payment is delayed to a later date or to be paid in installments. The price and the style of payments must be agreed at the onset of the agreement, without interest payment.

(Continued)

---

[1]Table 5.1 was constructed from the works of Muhammad al-Bashir (2001); Muhammad Rawwas Qal'ahji (2005); Mohd Ma'sum Billah (2006); Norman and Schmidt (2007); Islamic Development Bank (2007); Kamalpour (2006), etc.

Table 5.1. (*Continued*)

Product	Style of contract	Brief description
Al-Murabahah	Sale and purchase on a cost plus basis	A contract of sale and purchase whereby one party agrees to buy an asset from a third party (in the market) and then sells it to the other party at a price that includes his original purchased price, administrative costs and a reasonable profit. The second party pays the first party through an agreed style of payment usually deferred to a later date.
Al-Musharakah	Equity financing	This contract is similar to the *al-mudaraba* except that the *mudareb* provides part of the equity. Profits and losses are shared between the parties.
Al-ijarah	Lease	In this contract the principal purchases and leases out an asset or equipment required by his client for an agreed rental fee. This fee can be fixed in advance or be subject to occasional reviews. During the rental period the leased asset remains under the control of the principal.
Salam	Advance purchase	The *salam* contract is defined as the advance purchase of specified goods with full pre-payment.
Ju'alah	Some sort of a reward contract	For example a contract for bringing back fugitives or lost property. The contract is binding once the object is realized.
Istisna' or *Al-Muqawalah*	Manufacturing contract	A contract to make or produce something of which the subject matter is clearly specified and the terms of payment agreed at the onset.
Takaful	Insurance	A contract on the provision of insurance that is Shariah-compliant.
Sukuk	Bonds	Asset backed securities issued in line with the Shariah.

Source: Khairuddin (2009), p. 109.

Current Contract Practice for Construction Works

The business of construction is unique and complex. It involves elaborate commissioning, design, management and assembly of resources — human, finance, materials, plant and equipment — over a period of time (Khairuddin, 2002).

Contracts are entered by the parties to provide evidence that a legal relationship exists between them; it outlines the rights, duties and responsibilities of each party in relation to the works being undertaken and describes, among others, how the works are to be performed, payments to be made and disputes to be resolved.

Currently, construction contracts are modeled after the conventional-styled contractual practices. They are administered in accordance with a country's statutes on contractual relationships, common law, customs and traditions.[2] A typical construction contract has many characteristics including:

Firstly, contracts are often always awarded to contractors through complex systems of tendering and procurement. The systems entail either:

(1) The client provides complete information including drawings, bills of quantities and specifications of the proposed works and the contractor provides construction services as practiced under the design-bid-build (DBB) system of procurement;

(2) The client provides only brief requirements while the successful contractor provides complete design and construction services as practiced under the design-build or turnkey (DB/T) system of procurement;

(3) The management approaches system of procurement whereby a professional manager or a contractor provides complete management services and works' contractors perform the actual tasks of construction; or

---

[2] As an example, in Malaysia a construction contract is a branch of private law. It is governed under the Contracts Act 1950 and practiced in accordance with the customs and traditions of business transactions peculiar to Malaysia.

(4) Works may be procured under any one of the newer methods of project delivery systems collectively referred to as the Public Private Partnership (PPP) such as Privatization and the Private Finance Initiative (PFI).[3]

Secondly, there exist a myriad of parties and activities that transcend the construction industry's supply chain: from initiation of the works to construction, completion and maintenance of the completed facility. Basically, the parties to a construction contract are the client and the main contractor. The architects, quantity surveyors, engineers, etc. are not parties to this contract. They have their own terms of employment with either the client or contractor. Therefore, in a construction project, a myriad of primary and secondary contractual relationships are present.

Thirdly, the contracting parties enter into a contract in which the subject matter is nonexistent at the time of the contract. Details of the subject matter are described and illustrated in the contract documents. The key contract documents are:[4]

(a) Articles of Agreement
(b) Form of Tender (where tenderers submit their offer)
(c) Letter of Acceptance (where the client informs the successful tenderer that his offer is being accepted)
(d) Conditions of Contract and Its Appendices
(e) Special Provisions to the Conditions of Contract
(f) Contract Drawings
(g) Bills of Quantities or Schedule of Rates and Summary of Tender Specifications
(h) Other relevant information listed in the appendix to the Conditions of Contract.

Fourthly, there are alternatives to the way in which the price of a contract is fixed. The more commonly used methods are either

---

[3]The author acknowledges that there are many sub-systems within the PPP family including the Build-Operate-Transfer (BOT), Build-Transfer-Operate (BTO), Build-Own-Operate-Transfer (BOOT), Build-Lease-Transfer (BLT) etc.
[4]The list refers to documents found in a typical public works contract.

the lump sum measured contract based on bills of quantities; lump sum contract based on drawings and specifications; or reimbursable or cost plus contracts.

Finally, the practice of contracting has always been to use the commonly available and construction industry specific standard forms of contract. The dominant standard forms of contract used in Malaysia are the Public Works Department (PWD) 203 series for public works, the Persatuan Akitek Malaysia (PAM) forms for private building works and the Construction Industry Development Board (CIDB) forms.

A simple examination of the standard forms of contracts mentioned above was carried out to determine the presence or otherwise of the basic principles of *al-muamalat* mentioned hereinbefore. The outcome of the examination indicates that several basic principles of *al-muamalat* are absent (Table 5.2). Specifically, the basic principles that are absent or not expressly denounced are: *al-ihtikar* and *talaqqi al-rukban*, *riba*, trading in alcohol, gambling, prostitution, *gharar* and *maysir*. Coincidently, these are matters that are strictly prohibited by the Shariah. Thus, a contract that failed to denounce or forbid the presence of these matters are deemed as a non-Shariah-compliant contract.

Khairuddin (2009, pp. 112–113) identified the key differences between a Shariah-compliant contract and the conventional-styled contract to be in the area of faith or one's belief system. In other words, validity of a contract, from the perspective of the Shariah, is one that is subject to divine sanction; the subject matter and consideration for the contract must be in conformity with divine laws; the presence of usury or bribery renders the contract void and contracting is considered part and parcel of the religious belief and practices of Muslims.

Shariah-Compliant Construction Contracts: The *Istisna'*

One key feature of a construction contract is that the contracting parties enter into an agreement in which the subject matter is non-existent at the time of the contract. Instead, the subject matter is as described or illustrated in the contract documents.

Table 5.2.   Comparison between Shariah-Compliant Contract and Conventional-Styled Contracts.

	Basic principles of *al-muamalat*	Contracts	
		Shariah-compliant	Conventional
1	Business transactions are not forbidden	√	√
2	Contract is to facilitate business transactions	√	√
3	*Iktinaz* and *talaqqi al-rukban* leading to deception or inflated prices of goods or commodities are forbidden	√	X
4	*Riba* (interest charges), trading in alcohol, gambling, prostitution, etc. are forbidden	√	X
5	Avoidance of monopoly	√	√
6	Business transactions should be conducted with patience and tolerance	√	√
7	Uphold the concepts of truthfulness, ethics, integrity and accountability at all times	√	√
8	*Gharar* and *maysir* are forbidden	√	X
9	Parties in a contract must fulfil their respective obligations	√	√
10	Be diligent and persevere in conducting the businesses	√	√
	Elements of the *al-aqad* (contract)	Shariah-compliant	Conventional
11	*Sighrah*: *Ijab* and *qabul* (offer and acceptance)	√	√
12	*Aqidan* (qualifications of the contracting parties)	√	√
13	*Ma'aqud Alaid* (subject matter and consideration)	√	√

However, a Shariah-compliant contract requires physical presence of the subject matter. This is to allow the contracting parties to examine the subject matter and satisfy themselves prior to entering into the contract.

In this respect Muslim scholars and others concur that the *Istisna'* or *Al-Muqawalah* is considered the most appropriate

Shariah-compliant contract agreement to be used[5] (Muhammad Al-Amine, 2001; Muhammad Rawwas, 2005; Parker, 2006; Al-Baraka Banking Group, n.d).

In Arabic the term *Istisna'* means "making, manufacturing, or constructing something." An *Istisna'* contract is therefore "a contract with a manufacturer to make something..." (Muhammad Al-Amine, 2001, pp. 6–7).

Istisna' as a term denoting the act of making or manufacturing can be found in the Quran and *Sunnah* i.e. *"...(Such is) the work of Allah, Who has rightly perfected all things..."* (Quran 27: 88); and it was reported that the Prophet Muhammad (*s.w.t*) had ordered a craftsman to make a ring for him. In addition, the *Istisna'* has a long history as it was practiced by the Arabs during the pre-Islamic era (Muhammad Al-Amine, 2001, p. 6).

The key conditions for the legality of the *Istisna'* includes (Muhammad Al-Amine, 2001; Muhammad Rawwas, 2006; Al-Baraka Banking Group, n.d):

(1) General principles of *al-muamalat* are applicable in an *Istisna'* contract.
(2) For a contract to be valid:

   a. The contracting persons must possess all the qualifications (of a sound mind and has reached puberty, is free to enter into a contract and not a bankrupt);
   b. The process of offer (*ijab*) and acceptance (*qabul*) takes place; and
   c. The subject matter and the consideration is of value, is certain (*gharar* or uncertainty is prohibited) and one that is not prohibited by the Shariah.

---

[5]For detailed discussions on the *Istisna'*: its concept, legality, relationship with other contracts i.e. *salam, ijarah, ju'ala, murabahah* and sale and on how the four schools of Islamic scholars view *Istisna'* see for example Muhammad al-Amine (2001).

(3) The specific conditions for the *Istisna'* are:

    a. The object, in this case the construction of the structure to be built, must be clearly defined, illustrated, quantified and specified so as to avoid ambiguity or misrepresentation that could lead to dispute;

    b. The structure to be built should be things that people customarily deal with;

    c. The time of completion must be specified;

    d. The materials to be used in the works should be supplied by the contractor (if the materials are supplied by the client the contract is not *Istisna'* but *Ijarah*); and

    e. Advance payment is not a condition but is permissible or payment may be made in installments or at the end of the contract.

(4) In addition to the above, it is permissible in an *Istisna'* contract to include the following provisions:

    a. Liquidated and Ascertained Damages (LAD), being a prior agreement between the parties to a contract about a sum payable in the event of one party failing to complete or delaying performance of his obligations under the contract;

    b. Circumstantial changes and their effects such as exceptional events or unforeseeable events arising from general circumstance that occurred during the contract; and

    c. Dispute resolution through arbitration.

(5) An *Istisna'* contract is terminated once the contractor completes the project and hands it over to the client and the latter pays the former. However, termination during the progress of the works may take place upon the death of one of the parties (should the party who suffers death be a sole proprietor).

Available literature indicates that current practice of the *Istisna'* contracts for construction works focus on financing contracts by financial institutions. Box 5.1 provides a description on the concept of *Istisna'* practiced by the Islamic Development Bank (IDB). In such

"Definition

Istisna'a is a contract whereby a party undertakes to produce specific goods and services, and made according to certain agreed-upon specifications at a determined price and for a fixed date of delivery. The production of goods includes any process of manufacturing, construction, assembling or packaging. In *Istisna'a*, the work is not conditioned to be accomplished by the undertaking party alone, and this work or part of it can be done by others under his control and responsibility. *Istisna'a* could also be used in pre-shipment financing of the acquisition of capital goods. In addition the *Istisna'a* mode can be used to finance intangible goods such as gas and electricity for which Leasing or Installment Sale modes are not suitable.

Objectives

The main objective of the *Istisna'a* mode of financing is to promote manufacturing and construction capabilities in the IDB member countries. This may relate to manufacturing of complete assets in the form of capital goods or construction of certain infrastructure projects such as rail, roads, schools, bridges, buildings, etc. This mode of financing can also be applied to Export Financing Scheme (EFS) to enhance intra-trade among IDB member countries.

Scope and Eligibility

Istisna'a provides medium- to long-term financing to meet financing requirements for manufacturing/constructing/sup-plying/sale of identified goods and assets such as industrial or construction equipment, machinery, cargo vessels, oil tankers, trawlers, dredgers, locomotives, transport equipment, pipe-lines for water and oil distribution, gas and electricity and their transmission or distribution lines, electric generators and transformers, telecommunication equipment, oil rigs, hospital equipment, buildings, etc. Under the *Istisna'a* mode, it is also possible to finance intangible assets like gas, electricity, etc. Additionally, unlike under Leasing and Installment Sale, *Istisna'a* can be used to finance working capital. *Istisna'a* financing period is determined by the time required for procurement of the necessary materials and actual manufacture of the goods according to the agreed contract."

Box 5.1. *Istisna'* According to the Islamic Development Bank.
Source: IDB, 2007.

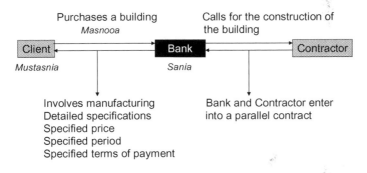

Figure 5.1. *Istisna'* Financing Contract.

cases, parallel *Istisna'* is being applied (see illustration in Figure 5.1). However, parallel *Istisna'* is contrary to the typical conventional-styled construction contract arrangement whereby the employer contracts direct with the contractor.

Review on Key Provisions of Current Contract Forms to Ascertain Their Compliance or Otherwise with the Shariah

In the construction industry each standard form of contract is a very comprehensive document and to many it reflects fairness to the contracting parties. The forms identify among others the rights and responsibilities, rewards or punishments, and allocate risks associated with the works among the contracting parties. However, what is fair or otherwise is open to the parties' own interpretation.

Standard forms of contracts

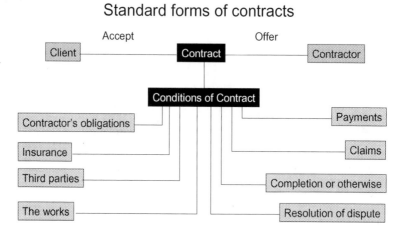

Figure 5.2.   Key Provisions of Typical Standard Forms of Contract.

Differences exist between the standard forms of contract but overall there is consistency in terms of the key provisions in the conditions i.e. all standard forms of contract address key elements of the works, from start to completion (Figure 5.2).

A pilot study, based on review of literature, was carried out to ascertain whether the key provisions in the current and conventional-styled contract for construction works comply or otherwise with the Shariah. In the study, 10 key provisions contained in the PWD 203A standard form of contract, one of the dominant standard forms of contract in use by the Malaysian public sector, were selected to represent the conventional-styled construction contract practice and the corresponding principles of the Shariah and *Istisna'*. Initial findings were then presented and consensus achieved through a Focus Group Discussion (FGD) with the construction industry and Shariah experts. Table 5.3 provides a summary of the study's outcome.

Conclusion

This chapter discussed the concept and application of Shariah-compliant contract for construction works. Examination of the current and conventional-styled construction contract practices, focusing on the manner in which contracts were established and the

Table 5.3.   Comparison Between Conventional Contract Practice and the Shariah (*Istisna'*).

Conventional-styled contract practice (PWD 203A) as model	Shariah/*Istisna'*	Findings from the pilot study
1 **Formation of a contract**		
Capacity to contract. Generally, every person can enter into a contract on condition they are not infants or minors, not of an unsound mind or intoxicated. A company can enter into a contract on matters that are within its object clause.	The contracting persons must possess all the qualifications (be of a sound mind and reached puberty, are free to enter into a contract and not a bankrupt).	Broadly, the conventional practice appears consistent with the *Istisna'*.
Offer and acceptance. Prior to a contract there must be an offer made by the contractor, the offer made on the basis of detailed terms and conditions and follows stipulated procedures that include delivery of the offer on or before the closing date and time, on a special form, etc.	*Ijab* and *qabul*. The conclusion of a contract is a result of the connection of an offer (*ijab*) and an acceptance (*qabul*) during a contractual "session" called *majlis*.	Data on the practice of an *ijab* and *qabul* and the procedures they entail are currently not available.
A contract exists once an offer is formally accepted. The acceptance must match the terms of the offer. There are instances where clients resort to provide conditional offers or what is sometimes referred to as "Letter of Intent."	Oral contracts are enforceable depending upon proof of contract but written contracts are preferred as they clearly indicate the terms of the agreement between the parties (SAGIA, 2005).	

(Continued)

Table 5.3.   (*Continued*)

Conventional-styled contract practice (PWD 203A) as model	Shariah/*Istisna'*	Findings from the pilot study
Consideration. In construction contracts, the consideration of the contractor to fulfill his obligations under the contract is that of payment of a price by the client (Contract Sum). To make a consideration valid it must have value, not be against the law and have nothing to do with events in the past.	The subject matter and the consideration is of value, is certain (*gharar* or uncertainty is prohibited) and one that is not prohibited by the Shariah.	Broadly, the conventional-styled practice appears consistent with the *Istisna'*.

2 **Contractor's obligations**

The contractor has many duties to perform. However, his primary obligation is to construct and complete the works in accordance with the contract documents. In fulfilling his obligations the contractor shall provide everything necessary: materials, goods, skills, etc. The obligations may or may not include design and maintenance responsibilities (depending on the procurement system used).	The contractor promises to construct the building, etc. based upon clearly defined, illustrated, quantified and specified models so as to avoid ambiguity or misrepresentation that could lead to dispute. The structure to be built should be things that people customarily deal with and the time of completion specified.	Broadly, the conventional practice appears consistent with the *Istisna'*.

(*Continued*)

Table 5.3.   (*Continued*)

Conventional-styled contract practice (PWD 203A) as model	Shariah/*Istisna'*	Findings from the pilot study
The contractor shall proceed with the works regularly and diligently from the date for possession of the site and to complete before or on the date for completion as stated in the contract.	Materials to be used in the works should be supplied by the contractor (if the materials are supplied by the client the contract is not *Istisna'* but *Ijara*).	
A Superintending Officer (S.O) is employed by the client to provide overall supervision and direction to the works.	In *Istisna'* the contractor may perform all the works himself or he may appoint others to work on his behalf. However, if the client stipulates that the works shall be manufactured by the contractor or a specific manufacturer or manufactured from specific materials then the contractor must adhere to such requirement. The contractor is liable for any defects to the works; this principle differentiates between a *ribawi* investment and a Shariah-compliant investment (Muhammad Al-Amine, 2001, pp. 47, 53).	

(*Continued*)

Table 5.3. *(Continued)*

Conventional-styled contract practice (PWD 203A) as model	Shariah/*Istisna'*	Findings from the pilot study
3 Insurances		
Insurance against injury to persons and damages to property. The Contractor is liable and must indemnify the client against any expense, liability, loss, claims or proceedings at statute or common law for personal injury or death and injury or damage to property due to the carrying out of the works. Consequently, the contractor must maintain insurances in respect of the liabilities.	There are Shariah-compliant insurances under the various *takaful* schemes.	Conventional insurance is *haram*. To be Shariah-compliant the contract should include provisions making *takaful* mandatory.
Insurance of the works. The contractor must insure the works against specified risks (fire, floods, etc.) for the full reinstatement value of all works executed, materials on site, etc.		
Workmen compensation/SOCSO. Similar to above but specifically for workmen employed in and for the performance of the contract.		

(Continued)

Table 5.3.   (*Continued*)

Conventional-styled contract practice (PWD 203A) as model	Shariah/*Istisna'*	Findings from the pilot study
4 Third parties		
The contractor is allowed to sub-let and/or assign the works to others on condition he obtains approval from the S.O and the client respectively.	In *Istisna'* the work is not conditioned to be accomplished by the undertaking party alone, as the works or part of it can be done by others under his control and responsibility.	Broadly, the conventional-styled practice appears consistent with *Istisna'*.
The client may nominate others as sub-contractors to be employed by the main contractor.	However, if the client stipulates the works shall be constructed by the contractor or a specific person or company then the contractor must comply by this requirement (Muhammad Al-Amine, 2001, p. 53).	
5 The works		
The contract should provide clear definition of the works: scope, start and completion dates, etc.	Similar to 2 above, i.e. contractor's obligations.	Broadly, the conventional-styled practice appears consistent with *Istisna'*.
6 Payments		
There are three types of payments to the contractor: advance payment, interim (usually monthly) payments and final payment.	Advance payment is not a condition in *Istisna'* but is permissible. Payments may be made in installments or are postponed until the works are completed. What is important is that the contract price should be determined.	Broadly, the conventional-styled practice appears consistent with *Istisna'*.

(*Continued*)

Table 5.3.   (*Continued*)

Conventional-styled contract practice (PWD 203A) as model	Shariah/*Istisna'*	Findings from the pilot study
Advance payment is not automatic but is contingent upon a set of conditions precedent.		

7 Damages for non-completion and LAD

If the contractor fails to complete the works by the date stated in the contract, the S.O must issue a certificate of non-completion and thereafter deduct from monies due to the contractor the liquidated and ascertained damages at the rate stated in the contract agreement.	It is permissible for an *Istisna'* contract to include a provision on liquidated damages. The provision refers to a prior agreement between the contracting parties about the sum to be paid in the event of the contractor failing to complete his contractual obligations within the time stated in the contract.	Broadly, the conventional-styled practice appears consistent with *Istisna'*.

8 Delay and extension of time

The contractor shall inform the S.O if there are reasons (confined to the reasons listed in the form) affecting progress of the works. The S.O may give to the contractor an extension of time for completion of the works.	Under the concept of *al-Jai'hah* change in circumstances have consequences. To be valid such event should be exceptional in nature, unforeseeable, widespread, and occurs during performance of the contract. The contracting parties may re-negotiate terms.	Broadly, the conventional-styled practice appears consistent with *Istisna'*.

Table 5.3.   (*Continued*)

Conventional-styled contract practice (PWD 203A) as model	Shariah/*Istisna'*	Findings from the pilot study
9 **Alternative method of dispute resolution through arbitration**		
Any dispute between the client and the contractor must be referred to the S.O for his decision failing which the parties may bring the matter to arbitration.	According to Muhammad Al-Amine (2001, p. 75) under the concept of *maslahah* arbitration is permissible and can be included in the contract.	Broadly, the conventional-styled practice appears consistent with *Istisna'*. However, Shariah-compliant ADR (*tahkim*) is preferred.
10 **When is a contractor discharged from his liabilities?**		
Discharge by execution of the contract. Discharge by rescission. Discharged by mutual consent.	Similar principles apply (Ahmad, 2008).	Broadly, the conventional-styled practice appears consistent with *Istisna'*.

key provisions of the conditions thereto, suggests that the current and conventional-styled construction contract practices appear to comply with the Shariah. The areas of non-compliance identified concerns the absence of specific provisions addressing key Shariah matters that are prohibited in Islam namely: *al-ihtikar* and *talaqqi al-rukban*, *riba*, trading in alcohol, gambling, prostitution, *gharar* and *maysir* and the application of conventional insurance instead of *takaful*.

Demand for Shariah-compliant construction contracts is set to grow. In addition, it is contended that Shariah-compliant construction contracts could provide a credible alternative to the current and conventional-styled construction contract on the basis that it emphasizes the concepts of justice, moral obligations, accountability, equality and its practice is seen by Muslims as an act of worship.

Given that *al-muamalat* in construction business transactions is still in its infancy, experts in Shariah, law and construction should increase their efforts to develop the concept further and promote its application.

References

Ahmad, MU (2008). Dischargeability and frustration with special reference to construction contract formation. In *Proc. Conf. Shariah Compliant Construction Contract*, International Islamic University Malaysia. Kuala Lumpur: April 16, 2008.

Al-Baraka Banking Group. (n.d). *Shariah Opinion (Fatwa) on Istisna', Contracting and Salam.* http://www.albaraka.com/media/pdf/Research-Studies/RSIS-200706201-EN.pdf [Accessed September 25, 2017].

Government of Malaysia (2010). Standard Form of Contract to be Used Where Bills of Quantities Form Part of the Contract. Public Works Department (PWD). Form 203A (Rev. 1/2010).

Government of Malaysia (2010). Standard Form of Contract to be Used Where Drawings and Specifications Form Part of the Contract. Public Works Department (PWD). Form 203 (Rev. 1/2010).

Islamic Development Bank (2007). *Modes of Project Finance.* www.isdb.org [Accessed April 3, 2007].

Kamalpour, A (2006). Islamic Finance — The Ijara Sukuk and other Shariah-compliant structures. *Dechert On Point*, Summer 2006(6), 4–5, www.dechert.com [Accessed July 4, 2007].

Khairuddin, AR (2002). *Construction Procurement in Malaysia. Processes and Systems, Constraints and Strategies.* Kuala Lumpur: International Islamic University Malaysia.

Khairuddin, AR (2007). Shariah compliant contract: A new paradigm in multi-national joint venture for construction works. In *International Joint Ventures: Reaching Strategic Goal*, K Kobayashi *et al.* (eds.), pp. 14–25. Bangkok: Asian Institute of Technology Thailand and Kyoto University Japan.

Khairuddin, AR (2009). Shariah compliant contract: A new paradigm in multi-national joint venture for construction works. In *Joint Ventures in Construction*, K Kobayashi, *et al.* (eds.), pp. 103–114. London: Thomas Telford.

Malaysia Institute of Architects (Pertubuhan Arkitek Malaysia — PAM) (2006). Agreement and Conditions of PAM Contract 2006 (With Quantities).

Malaysia Institute of Architects (Pertubuhan Arkitek Malaysia — PAM) (2006). Agreement and Conditions of PAM Contract 2006 (Without Quantities).

Mohd. Ma'sum, B (2006). *Shar'iah Standard of Business Contract.* Kuala Lumpur: A.S. Noordeen.

Muhammad Al-Amine (2001). *Istisna' (Manufacturing Contract) in Islamic Banking and Finance*. Kuala Lumpur: A.S. Noordeen.

Muhammad Rawwas, Q (2005). Translated by Basri bin Ibrahim. *Urusan Kewangan Semasa Menurut Perspektif Syariah Islam*. Kuala Lumpur: Al-Hidayah.

Norman, WC and SB Schmidt (2007). *Principles of Shari'a in Islamic Finance and Their Application for Western Investor*. Washington: Patton Boggs, LLP online [Accessed April 3, 2007].

Parker, M (2006). Istisna construction financing gains momentum. *Arab News*, January 23, 2006. www.arabnews.com [Accessed April 3, 2007].

SAGIA (2005). *Legal Guidelines. Saudi Arabian General Investment Authority*. www.sagia.gov.sa/ [Accessed March 4, 2007].

Chapter 6

Istisna' Model for Construction Works' Contracts

KHAIRUDDIN Abdul Rashid

Introduction

One commonly accepted Shariah compliance tool for business transactions that permits a contract to be formed, although the subject matter for the transaction is yet to exist, is *Istisna'*. However, the current application of *Istisna'* in construction works contracts appears to be motivated largely by the need to facilitate financing for the procuring party or the Employer instead of the *Istisna'* serving as the primary contract between the Employer and the Contractor. In other words, *Istisna'* has largely been turned into a Shariah-compliant financing tool. This situation, not to be viewed negatively, is hardly surprising as it is the Islamic financial institutions that have been actively promoting the use of *Istisna'*. Their motivation includes that *Istisna'* qualifies as another Shariah-compliant financial product that they are able to offer to their customers. As such and in order to facilitate the institutions' involvement a parallel *Istisna'* becomes a requisite.

The objective of this chapter is to critically analyze *Istisna'* for construction works contracts. The analysis is not on the legitimacy of *Istisna'* as this matter has been well established and documented but on the way in which *Istisna'* is being practiced *vis-à-vis* the conventional-styled construction works' contracts. The critical analysis is conducted via desk study and discussions with experts. For the purpose of setting direction to the analysis the following questions have been prepared:

(1) Is *Istisna'* the Shariah-compliant tool that contracting parties could use when entering into a construction works contract?
(2) How does *Istisna'* work?
(3) Is the current *Istisna'* practice consistent with the conventional construction works contracts?

The rest of the chapter is structured into six parts. Immediately after the introduction, the chapter presents a review on the current practice of conventional construction works contract, the purpose of which is to provide background information on the workings of such a contract, thus setting the tone for the ensuing discussions. This is followed by a review of literature on *Istisna'* and its application for construction works contracts. Part 4 presents the outcome of a critical analysis of the *Istisna'* as it is applied to construction works contracts. The penultimate part presents a proposed conceptual model of *Istisna'* for construction works contracts. Finally part 6 offers a conclusion addressing the directional questions prepared at the onset of the study and admissions of the study's limitations.

Conventional Construction Works Contracts

Essentially, the formation of conventional contracts, including those for construction works contracts, is governed by Act 136, Contract Act 1950. The practice is such that a works contract is almost always based on one of the currently available standard forms of contract either the PWD 203 or the PAM series.[1]

The formation and signing of a contract between the key contracting parties — the Employer and the Contractor — is preceded by elaborate processes of procurement that include the

---

[1] There are other standard forms of contract in use such as the CIDB form, etc.

processes of initiation, funding, design, costing, seeking statutory approvals, tendering for the Contractor's services, and selection and appointment of Contractor.

In construction works contracts it is the practice that the main parties to the contract namely the Employer and Contractor are the signatories to the contract. It is mandatory for the Contractor and others involved in the business of construction to be an entity registered with the Construction Industry Development Board Malaysia (CIDB Act 520). It is the responsibility of the Employer, among others, to provide to the Contractor all necessary information, supervision, instructions, approvals and make the necessary interim and final payments.

The Employer may engage advisors and consultants[2] under further contracts respectively. Detailed information on the contract is stipulated in the contract documents that include, among others, the terms and conditions of the contract (either the PWD 203 or PAM series), drawings, specifications and bills of quantities. The basis for the contract sum may either be a lump sum based on bills of quantities, lump sum based on drawings and specifications, or cost plus/reimbursable contracts. Payments to the Contractor are based on works done, in arrears as interim payments.

It is the responsibility of the Contractor to execute and complete the works all in accordance with the contract and in return to receive payments from the Employer. The Contractor may engage sub-contractors and suppliers to assist him with the works but he remains fully responsible to the Employer.

The Employer may enter into further contracts with financial institutions for the purpose of project finance. Likewise, the Contractor may enter into further contracts with financial institutions or creditors for the purpose of funding his construction activities as his payments are in arrears. The contractual relationships of the parties in a typical construction works contract are as illustrated in Figure 6.1.

---

[2]Usually the consultants are the Architects, Engineers and Quantity Surveyors.

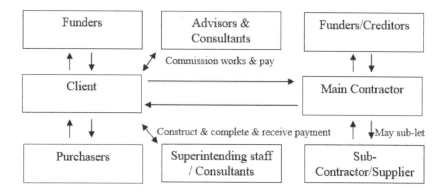

Figure 6.1.   Conventional-Traditional Construction Contract.

Istisna'

The legitimacy of *Istisna'* as a Shariah-compliant tool for entering into a business transaction in which the subject matter is yet to exist has been firmly established. The concept originated from the *hadīth* that Prophet Muhammad (pbuh) once requested for a golden ring to be manufactured for him (Bank Negara Malaysia, 2010). There is a consensus among the Muslim jurists on the use of the *Istisna'* (Bank Negara Malaysia, 2010) and that the International Islamic Academy of *Fiqh*[3] has also recognized the application of *Istisna'* (Decision no. 66/3/7 dated May 14, 1992 as per Box 6.1).[4]

According to the Islamic Development Bank (IDB, 2001), *"Istisna'* is a contract whereby a party undertakes to produce a specific thing which is possible to be made according to certain agreed-upon specifications at a determined price and for a fixed date of delivery." In a somewhat broadly similar but simplified definition, the Bank Negara Malaysia (2010) defined *Istisna'* as "... a contract to manufacture, build or construct assets to be delivered to the purchaser."[5]

---

[3]International Islamic Academy of *Fiqh*, 7th Session, Decision no. 66/3/7 dated May 14, 1992.

[4]Syed Alwi Mohamed Sultan (2007).

[5]See also definition by Bank Negara Malaysia (2015, p. 5) i.e. *"Istisna'* refers to a contract which a seller sells to a purchaser an asset which is yet to be constructed,

- *The contract of Istisna' is a binding contract on both parties if it satisfies all terms and conditions;*
- *The validity of Istisna' depends on:*
  - *The nature, quality, quantity, and the description of the asset to be manufactured is known*
  - *A time is fixed for manufacturing the asset.*
- *It is permissible in a contract of Istisna' to defer the entire payment or to pay by installments within a fixed time.*
- *It is permissible to include in a contract of Istisna' a clause about liquidated damages, if the parties so agreed, save for cases of force majeure or unforeseen events.*

Box 6.1.   International Islamic Academy of *Fiqh*'s Decision on *Istisna'*.
Source: Syed Alwi Mohamed Sultan, 2007.

The Islamic Development Bank states that *Istisna'* does not require the party undertaking to produce the "specific thing" to perform the works themselves as works may be performed by others who have control and responsibility (IDB, 2001). In other words, the manufacturer or contractor may sub-let the works to someone else. However, Bank Negara Malaysia (2010) implied that sub-letting is permitted if there is no express requirement from the procurer for the undertaking party to perform the works themselves.

Consequently, the *Istisna'* may be used as a Shariah-compliant tool by parties when entering into business transactions involving producing or manufacturing a specific product or asset such as constructing a building and the likes as long as the requirements (see Box 6.1) are satisfied.

The contractual relationships of the parties — Procurer-Contractor (and Sub-Contractor, if applicable) — in a typical but overly simplified *Istisna'* is as illustrated by Figure 6.2 and to be read together with Box 6.1 above.

---

built or manufactured according to agreed specifications and delivered on an agreed specified future date at an agreed predetermined price."

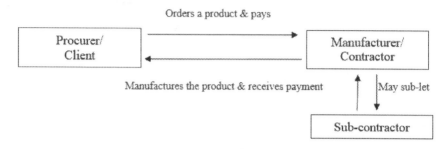

Figure 6.2.   Basic *Istisna'* Contract.

However, a review of currently available literature failed to identify construction works contracts that apply the basic or direct *Istisna'* contract between the Employer and the Contractor. What was found are two similar but distinct *Istisna'* models that have been structured so as to facilitate Islamic financial institutions to provide funding facility to the Employers. The situation arises because constructing a built asset such as a building involves among others, a considerable sum of money in which most, if not all, Employers would not be able to afford funding the project themselves. Instead Employers would enlist the help, in the form of loans, from Islamic financial institutions.

Istisna' *Contract Model 1*

Under this model an Employer and an Islamic financial institution enter into an *Istisna'* contract. Under the contract the latter is to undertake the construction and completion of the proposed built asset for the former and to provide funding facility. However, the Islamic financial institution has no business or expertise to construct and complete the built asset so they end up entering into an agency or sub-contract agreement with a Contractor. It is the Contractor that is being tasked to construct and complete the built asset for the Islamic financial institution. In this instance, the Contractor acted as an agent or sub-contractor to the Islamic financial institution. The model is similar to the one illustrated in Figure 6.2 with one notable exception i.e. the Islamic financial institution undertakes the responsibility but is not the manufacturer or constructor of the

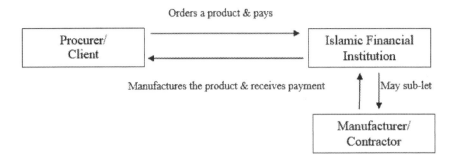

Figure 6.3. *Istisna'* Contract Model 1.

built asset. The contractual relationships for *Istisna'* model 1 is as illustrated by Figure 6.3.

Istisna' *Contract Model 2 — The Parallel* Istisna'

Under this model an Employer and an Islamic financial institution enter into an *Istisna'* contract (*Istisna'* contract 1). The Islamic financial institution, instead of acting as the undertaking party in the *Istisna'* contract, would enter into another *Istisna'* contract, this time with a Contractor (*Istisna'* contract 2). In the second *Istisna'* contract, the Contractor would undertake to construct and complete the built asset for the Islamic financial institution. Upon completion of the built asset the Contractor delivers it to the Islamic financial institution (*Istisna'* contract 2) and subsequently the Islamic financial institution delivers the same built asset to the Employer (*Istisna'* contract 1).

The key difference between the *Istisna'* contract models 1 and 2 is that in the second *Istisna'* model, the Islamic financial institution would not be appointing an agent (a sub-contractor) to construct and complete the built asset, instead they would, through another but parallel *Istisna'* contract, commission a Contractor to execute the works. The Contractor, in the normal cause of his business may sublet the works to a sub-contractor. This second *Istisna'* model wherein two independent *Istisna'* exist in a business venture i.e. to construct a built facility is also known as the Parallel *Istisna'*. The contractual relationships in a Parallel *Istisna'* is as illustrated in Figure 6.4.

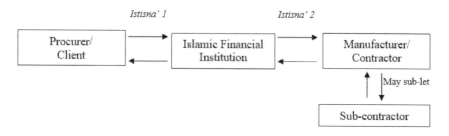

Figure 6.4. *Istisna'* Contract Model 2 — The Parallel *Istisna'*.

Critical Analysis of *Istisna'* and Conventional Construction Works Contracts

This part of the chapter presents a critical analysis of *Istisna'* for construction works contracts, the key objective being to assess its compatibility with the current practice of conventional construction works contracting.

Acceptability of The Istisna' for Use in Construction Works Contracts

The legitimacy of *Istisna'* as a Shariah-compliant tool for entering into a business transaction in which the subject matter is yet to exist (including for construction and completion of built assets) has been firmly established. There is consensus among the Muslim jurists on the matter and detailed workings of the *Istisna'* has been published by notable institutions including the IDB (2001), Bank Negara Malaysia (2010, 2015) and International Islamic Academy of *Fiqh*[6] (1992).

The Workings of The Istisna' and Conventional Construction Works Contract

The current study has identified three *Istisna'* models namely the Basic *Istisna'*, *Istisna'* model 1 and *Istisna'* model 2 also known as the Parallel *Istisna'*.

---

[6]Ibid at note 3.

From the literature review, it appears that the Parallel *Istisna'* is more popular and is often used by Islamic financial institutions to facilitate providing funding facilities to the procurers or Employers. As a consequent, *Istisna'* has been largely turned into a Shariah-compliant financing tool. The situation is hardly surprising as it is the Islamic financial institutions that have been actively promoting the use of *Istisna'*, their motivation includes that *Istisna'* qualifies as another Shariah-compliant financial product that they are able to offer to their customers. In order to facilitate the institutions' involvement, Parallel *Istisna'* becomes a requisite.

Detailed evaluation of the *Istisna'* models revealed that the Basic *Istisna'*, in terms of the presence of direct contractual relationship between the Employer and the Contractor, is consistent with the currently practiced conventional construction works contract.

In *Istisna'* models 1 and 2 the presence of the Islamic financial institution, acting as the middle-person, appears to not be the style commonly adopted in the conventional-styled construction works contracts. In fact, observing it from the perspective of the conventional way of construction works contracting, the presence of the Islamic financial institution acting as the middle-person, while appreciating that such a style stems from the need to facilitate the provision of a Shariah-compliant financing facility, may, in one way or another, provide unnecessary complications such as:

- The Islamic financial institution has no business therein apart from providing financing to the Employer. In addition, construction is not their core business and they probably possess little or no skills and expertise in construction;
- The presence of the Islamic financial institution, acting as the middle-person, would add another layer to the already complex processes of construction procurement. Almost all modern construction procurement systems such as Design and Build (D&B), Turnkey and the management-oriented procurement systems advocate direct contractual relationship or even single-point responsibility between the two contracting parties, i.e. the Employer and the Contractor. There is a lot to be gained

from having direct contractual relationship or even single-point responsibility including better and speedier communications, lesser adversarial situations leading to lower potentials for disputes between the contracting parties and down-time and therefore reducing greatly the potentials for cost and time overruns. Having a middle-person and therefore putting an extra layer of bureaucracy between the two contracting parties seems to be taking a step backward, insofar as the modern procurement system is concerned;

- In businesses, having a middle-person between two contracting parties could lead to additional costs but not necessarily value to the parties and the works;

- The current *Istisna'* arrangements appear to be focusing only on creating a pathway in facilitating achieving Shariah-compliant financing instrument for the Employer. However, there is more to it insofar as construction works contracts are concerned. As illustrated in Figure 6.1 there exist several other contractual arrangements in a particular construction works contract. These contractual arrangements may or may not apply provisions that are in compliance with the Shariah as studies by Khairuddin (2007), Siti Nora Haryati and Khairuddin (2009) and Khairuddin (2009) had found. Experts in Shariah-compliant financing tools and products should consider reviewing the entire contractual arrangements and adopt a more holistic approach rather than focusing only on the financing part and being opaque on the rest of the contractual arrangements;

- In Malaysia, an entity wishing to undertake construction works and acting as a contractor must be registered, in the appropriate and approved Grade, with the Construction Industry Development Board Malaysia (CIDB Act 520, Part VI Registration of Contractors). In looking at the *Istisna'* model 1, it appears that the Islamic financial institution may stand accused of violating Act 520 unless it secured registration as a Contractor with the CIDB. In the case of the Parallel *Istisna'* a similar situation may exist unless the terms of the contract provisions state clearly that the Islamic financial institution is not identified as the party having the responsibility to undertake the construction and completion of the built asset.

Proposed *Istisna'* Model for Construction Works Contracts

The analysis presented in the foregoing section suggests that there is a need for a more holistic *Istisna'* model for construction works contracts i.e. one that must look beyond the financing aspect of a construction contract, as what the current *Istisna'* models 1 and 2 appear to be.

A more holistic *Istisna'* model must include Shariah-compliant provisions at all levels: the arrangements and provisions in the primary contract between the Employer and the Contractors, the upstream contracts between the Employer, his team of advisors, financiers and purchasers of assets developed by the Employer (if applicable) respectively and the downstream contracts between the Contractor, his sub-contractors, suppliers, creditors and financiers respectively. Given the myriad of business transactions and activities therein, the potential for the provisions and/or activities to be not in compliance with the Shariah could be high particularly those that relate to *gharar*, *riba* and *maysir*.

In addition, the *Istisna'* model should ideally be structured so that bureaucratic levels are reduced to the very minimum.

Furthermore, parties to a contract should consider adopting greater transparency and accountability in their business dealings including adopting the concept of *Ta'awun* (Partnering) and alternative dispute resolution techniques such as *tahkim* (Arbitration) and *Sulh* (Consultation).

An initial idea on the various elements, components considered necessary in the development of a more holistic *Istisna'* for construction works contracts has been crystallized in the form of a conceptual model and is presented in Box 6.2.

Conclusion

At the onset of this chapter three questions related to *Istisna'* for construction works contracts were asked. The questions and their

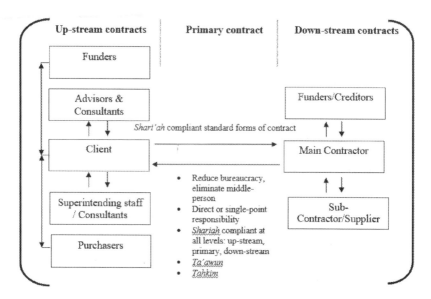

Box 6.2.   Proposed Model of a More Holistic *Istisna'* for Construction Works Contracts.

answers are:

1. Is *Istisna'* the Shariah-compliant tool that contracting parties could use when entering into a construction works contract?

The answer to this question is yes. Discussions on the legitimacy of the *Istisna'* have been presented in this chapter.

2. How does *Istisna'* work?

The workings of the current *Istisna'* may be categorized into three similar but distinct models: Basic *Istisna'*, *Istisna'* model 1 and *Istisna'* model 2 or Parallel *Istisna'*. It appears that the *Istisna'* models currently being practiced are models 1 and 2 with model 2 being most popular. The application of *Istisna'* appears to be more of a tool for Shariah-compliant project funding.

3. Is the *Istisna'* practice consistent with the current practice in conventional construction works contract?

The *Istisna'* practice is not consistent with the current practice in conventional construction works contract. Discussions on the identified areas of inconsistencies have been presented in this chapter. Some of the practices may not necessarily add value to the works or to the contracting parties instead as the arrangements may cause complications to the already complex processes of construction procurement.

Islamic financial institutions promote the use of *Istisna'* as a financing product that is Shariah-compliant whereas key players of the construction industry are keen to establish more holistic contractual arrangements that are Shariah-compliant, not just financing of the project alone. In attempting to address the desire of the latter a conceptual model of a more holistic *Istisna'* for construction works contracts has been developed. The model is herein presented for further deliberations.

Legislation

Contract Act 1950 (Act 136)
Construction Industry Development Board Act (CIDB Act 520)

References

Bank Negara Malaysia (2015). *Istisna'*. Kuala Lumpur: Bank Negara Malaysia.

Bank Negara Malaysia (2010). *Draft Shariah Parameter Reference 5: Istisna Contract*. Kuala Lumpur: Bank Negara Malaysia.

Government of Malaysia (2010). Standard Form of Contract to be Used Where Bills of Quantities Form Part of the Contract. Public Works Department (PWD). Form 203A (Rev. 1/2010).

International Islamic Academy of *Fiqh*, 7th Session, Decision no. 66/3/7 dated May 14, 1992.

Khairuddin, AR (2007). Shariah compliant contract: A new paradigm inmulti-national joint venture for construction works. In *International Joint Ventures: Reaching Strategic Goals*, K Kobayashi, *et al.* (eds.), pp. 14–25. Bangkok: Kyoto University and Asian Institute of Technology.

Khairuddin, AR (2009). Shariah compliant contract: A new paradigm in multi-national joint venture for construction works. In *Joint Ventures in Construction*, K Kobayashi, *et al.* (eds.), pp. 103–114. London: Thomas Telford.

Malaysia Institute of Architects (Pertubuhan Arkitek Malaysia — PAM) (2006). Agreement and Conditions of PAM Contract 2006 (With Quantities).

Malaysia Institute of Architects (Pertubuhan Arkitek Malaysia — PAM) (2006). Agreement and Conditions of PAM Contract 2006 (Without Quantities).

Siti Nora Haryati, AH and AR Khairuddin (2009). A study to assess whether current contract practice for construction is in compliance with Shariah or otherwise (with specific reference to construction contract formation). In Khairuddin Abdul Rashid, Kobayashi, K., Omoto, T., Preece, C.N. and Sharina Farihah Hasan, (Eds.). *Collaborative Efforts in International Construction Management*. IIUM, Kyoto University & Scholar Mind publishing.

Syed Alwi Mohamed Sultan (2007). *A Mini Guide to Shariah Audit for Islamic Financial Institutions — A Primer*. CERT Publications.

The Islamic Development Bank (2001). *Istisna'a Mode of Financing*.

Chapter 7

Validity of *Istisna'* for Construction Works Contracts

KHAIRUDDIN Abdul Rashid

Introduction

The application of *Istisna'* in construction works contracts is in its infancy. Consequently, there would be issues requiring critical assessment. In this chapter, an attempt is made to assess *Istisna'* from the viewpoint of its validity in construction works contracts.

This chapter is an extension of the earlier chapter titled *Istisna'* Model for Construction Works' Contract. Refer to Chapter 6 for a review of the *Istisna'* and conventional construction works contracts respectively.

Following this introduction a critical examination on the issues of validity of the *Istisna'* as it is being applied for construction works contracts will be presented. The focus of the examination is on the validity of *Istisna'* in terms of how the works are being presented in a contract and time or contract period. In the former, the discussion will include a review of literature on Bills of Quantities (BQ). Finally, the chapter offers a conclusion with recommendations on how best to bring forward the issue on validity of *Istisna'* in construction works contracts.

Istisna' and Its Applicability in Construction Works Contracts

Khairuddin (2009) and Khairuddin and Sharina Farihah (2014) studied *Istisna'* as promoted by the International Islamic Academy of *Fiqh* (1992), IDB (2001) and Bank Negara Malaysia (2010, 2015), among others. It appears that the ideas promoted by these three institutions on the *Istisna'* are generally consistent with each other. Their ideas focus principally on facilitating and legitimizing Shariah-compliant project financing. This situation arises due to the costs of constructing a built asset such as a building involving a considerable sum of money in which most, if not all, Employers would not be able to afford to fund the project themselves. Instead Employers would enlist help, in the form of loans, from Islamic financial institutions.

In this chapter, an attempt is made to critically examine *Istisna'* from the viewpoint of its validity in construction contracts. The approach is taken by way of examining the validity conditions as proposed by the International Islamic Academy of *Fiqh* (1992) (refer to Chapter 6, Box 6.1).

Applicability of Istisna' for Construction Works Contract, Issue 1: Validity of Contract — Descriptions of The Works

The International Islamic Academy of *Fiqh* (1992) has stipulated the conditions on the validity of an *Istisna'* contract.

The validity of *Istisna'* depends on:

○ The nature, quality, quantity and the description of the asset to be manufactured is known

On the above condition it would appear that the conventional construction works contract, with the requirements, in terms of the Drawings, Specifications and BQs forming part of the tender and contract documents (PWD 203A (Rev. 1/2010), Clause 1.1(b)), enable the nature, quality and description of the building to be constructed to be known clearly to all parties in advance of the works being done. In fact, with modern technology, the matter is soon going to be clearer when three-dimensional modelling, Virtual Reality and Building Information Modelling (BIM) become the norm in construction documentation. Consequently, it appears that the conventional construction works contract fulfills the said validity requirement.

However, the same cannot be said in *Istisna'* for construction works contract. Extensive literature review has failed to identify how such requirement is to be fulfilled: nothing on preparation, format, standard, methodology and detailing, documentation and presentation has been prescribed.

Thus, it is recommended that *Istisna'* for construction works contracts should adopt the strict regime on the requirements for descriptions of the works to be known in advance, as it is being practiced in conventional construction works contracts. Thus, all *Istisna'* contracts should have the Drawings, Specifications and BQs forming part of the contract.

In addition, it is strongly recommended that BQs should be made compulsory for all contracts, conventional and *Istisna'* alike. The presence of the BQs would enable parties to be aware, specifically, of their risks and responsibilities in detail. Furthermore, the presence of the BQs would facilitate minimizing the potential presence of *gharar* and *maysir* in construction works contracts. Thus:

> "O believers! Do not consume (use) your wealth among yourselves illegally (such as means of cheating, gambling and others of illegal nature), but rather trade with it by mutual consent..." (Quran 4: 29).

> "O believers! Wine and gambling, idols, and divining arrows are (all of them) abominations devised by Satan. Avoid them, so that you may prosper." (Quran 5: 90).

> "... and give full measure and weight — in justice... " (Quran 6: 152).

> "... and fulfill every covenant (to Allah and men). Surely every covenant will be inquired into." (Quran 17: 34).

> "And give just weigh and fall not short in the balance." (Quran 55: 9).

Bills of Quantities

In a typical contract a Bills of Quantities (BQ) is perhaps the most significant and elaborate document. Basically, it is a document that lists down, among others, the requirements of the works in terms of labor, plant, materials, quantities and quality of workmanship.

The term *quantities* refer to the amount of labor and materials required in the execution of the various items of the works. Together these items give the total requirements of the contract. This aspect of the BQ is stressed in clause 26.1 of the P.W.D. Form 203A (Rev 1/2010), thus:

> The quality and quantity of the Works as set out in the Bills of Quantities shall be the basis of the Contract Sum.

In the BQ the works are presented in a logical sequence and recognized manner, in order that they may be readily priced by the contractors during tenders (Seeley, 1972). The items of work are measured in recognized units, the method of which are detailed in the relevant Standard Method of Measurement (SMM) of Building Works for building projects and in the Civil Engineering Standard Method of Measurement (CESMM) for civil engineering works.

The following are considered to be the principal functions of a BQ (Seeley, 1972; Turner, 1979):

1. To provide a uniform basis for contractors tendering for a job — this is the single most important function of the BQ. It enables all contractors tendering for a job to price on exactly the same information with minimum effort.

2. To provide a basis for the valuation of variations — variations often occur during the progress of the work. The unit rates entered by the contractor during tendering will be used in the valuation of similar items of work executed as variations, either for additions or omissions.

3. To provide construction management information for all parties in the contract — the detailed nature of the BQs provide valuable information on the design, specifications, details of materials and workmanship, positional and labor requirements for the job. The information helps the Contractor and the Employer in the programming and cost management of the project.

4. To provide basis for cost planning and cost analysis — the priced BQs provide valuable information to enable the Employer to

analyze his cost plan and for detailed comparison. Its main purpose is, in this respect, to provide data for future cost plan and the like.

There are generally two types of BQs: firm BQs and the approximate BQs. Both types are similar in all aspects. However in the former, quantities entered therein are firm quantities whereas in the latter the quantities are approximate in nature. The approximate BQs are used in situations where time is pressing that prevent for firm design and firm quantities to precede tendering. In these conditions, the approximate BQs are used at the contract stage. Complete re-measurement, based on the completed design and completed work, and complete re-pricing using the tendered unit rates to achieve the final account will have to be done. Under this method the contractor is responsible only for his price. Any errors or omissions in the BQs will be corrected and adjusted in the final account of the contract.

Bills of Quantities are widely used in project procurement in Malaysia (Bahri, 1986; Khairuddin, 2002). For instance, the Public Works Department (JKR), the agency responsible for implementing all government projects, uses the traditional procurement system based on BQs as its main method of procurement. This method satisfies the government procurement requirements that: (i) all projects must be let on open competition; (ii) each project must be clearly defined; and (iii) that all items of work are accounted for. Similarly, most Employers prefer to use BQs in the procurement of their construction projects.

Bills of Quantities are prepared by Quantity Surveyors principally from drawings, specification and other documents provided by other members of the design and project team. Figure 7.1 provides a sample page of a typical BQ.

Applicability of Istisna' for Construction Works Contract, Issue 2: Validity of Contract — Time

The International Islamic Academy of *Fiqh* (1992) has stipulated the conditions on the validity of an *Istisna'* contract.

				SECTION 'F4' - BUILDING WORKS - BUNGALOW (TYPE 'A')	
ITEM	DESCRIPTION	UNIT	QTY	RATE (RM)	AMOUNT (RM)
	ELEMENT NO. 1 - WORK BELOW LOWEST FLOOR FINISH				
	Excavate and get out, part return, fill in and ram, deposit and spread in making up levels where directed within the site and remainder load and cart away				
A	Pit for pile cap not exceeding 1.50m deep	M3	16	15.60	249.60
B	Pit for pile cap exceeding 1.50m but not exceeding 3.00m deep	M3	2	16.64	33.28
C	Trench for ground beam not exceeding 1.50m deep	M3	5	15.60	78.00
D	Oversite to reduce level average 250mm thick	M2	39	3.90	152.10
E	Ditto 200mm thick	M2	160	3.12	499.20

F	100mm Thick bed of hardcore as described spread, levelled, well rammed and consolidated, watered and blinded with fine sand well rolled in and finished to receive apron (measured separately)	M2	17	9.46	160.82
G	100mm Thick ditto pool deck (Provisional)	M2	39	9.46	368.94
	50mm Thick lean concrete Grade 15 as described blinding screed spread and levelled under				
H	Pile cap	M2	11	9.88	108.68
J	Ground beam	M2	13	9.88	128.44
K	Ground slab	M2	114	9.88	1,126.32
L	Apron	M2	17	9.88	167.96
M	Pool deck (Provisional)	M2	39	9.88	385.32
	Vibrated reinforced concrete Grade 25 as described filled into formwork and well packed around reinforcement in				
N	Pile cap	M3	8	190.01	1,520.08
P	Column stump (Provisional)	M3	3	190.01	570.03
Q	Ground beam	M3	5	190.01	950.05
R	150mm Thick concrete ground slab	M2	132	28.50	3,762.00
S	100mm Ditto	M2	11	19.03	209.33
T	100mm Thick apron	M2	17	19.03	323.51
U	150mm Thick pool deck (Provisional)	M2	39	28.50	1,111.50
			To Collection :		11,905.16

Figure 7.1.   A Sample Page of a Typical BQ.

Source: Courtesy of a quantity surveying firm.

The validity of *Istisna'* depends on:

o A time is fixed for manufacturing the asset

In conventional construction works contracts, the time from the commencement of each contract and its completion is clearly defined. The dates are duly inserted and agreed upon at the signing of the contract. In addition, mechanisms to cater for the unlikely event that extension of time is required is clearly provided in the contract (PWD Form 203A (Rev. 1/2010)). However, the same cannot be said for *Istisna'* construction works contract. Consequently, it is recommended that the conventional practice in determining the start and completion times of a construction works contract and extensions thereto be adopted into *Istisna'* construction works contracts (with the appropriate modification, if necessary).

> "O believers! When you contract a debt from one another for a fixed period, put it (its amount and period of repayment) in writing. And let a scribe write it down between you justly (truthfully), and no scribe should refuse to write as Allah has taught him. . . and call in two of your men as witnesses. . ." (Quran 2: 282).

Conclusion

This chapter examined the *Istisna'* from the viewpoint of its validity in construction works contracts. Two conditions on validity as proposed by the International Islamic Academy of *Fiqh* (1992) were critically examined and it was found that:

Validity of Contract, Condition 1 — Description of the Works

The validity of *Istisna'* depends on the nature, quality and quantity, with the description of the asset to be manufactured known. It appears that the conventional construction works contract, with the requirements — in terms of the Drawings, Specifications and BQs forming part of the tender and contract documents — meets the stipulated requirement. Consequently, it is recommended that *Istisna'* for construction works contracts should have the Drawings, Specifications and BQs forming part of the contract. The presence of the BQs would enable parties to be aware of their risks and responsibilities and would also facilitate minimizing the potential presence of *gharar* and *maysir* in construction works contracts.

Validity of Contract, Condition 2 — Time

Time, i.e. duration of the contract, from commencement to completion must be made clear to the contracting parties. In the case of conventional construction works contracts, the time from the commencement of each contract, its completion and extension thereto is clearly defined. Consequently, it is recommended that the conventional practice in determining the start and completion times of a construction works contracts and extensions thereto be adopted into *Istisna'* construction works contracts (with the appropriate modification, if necessary).

References

Bahri, MA (1986). Application of turnkey system on building projects. *Surveyor*, 4th Quarterly, 54–57.

Bank Negara Malaysia (2015). *Istisna'*. Kuala Lumpur: Bank Negara Malaysia.

Bank Negara Malaysia (2010). *Draft Shariah Parameter Reference 5: Istisna Contract*. Kuala Lumpur: Bank Negara Malaysia.

Government of Malaysia (2010). Standard Form of Contract to be Used Where Bills of Quantities Form Part of the Contract. Public Works Department (PWD) Form 203A (Rev. 1/2010).

International Islamic Academy of *Fiqh*, 7th Session, Decision no. 66/3/7 dated May 14, 1992.

Khairuddin, AR (2002). *Construction Procurement in Malaysia. Processes and Systems, Constraints and Startegies*. Kuala Lumpur: IIUM Press.

Khairuddin, AR (2009). Shariah compliant contract: A new paradigm in multinational joint venture for construction works. In *Joint Ventures in Construction*, K Kobayashi, *et al.* (eds.), p. 103–114. London: Thomas Telford.

Khairuddin, AR and H Sharina Farihah (2014). Proposed Istisna' model for construction works. *Proc.* 2nd Kyoto University-IIUM Research Colloquium, IIUM, May 8, 2014.

Seeley, IH (1972). *Building Quantities Explained*. London: The Macmillan Press.

The Islamic Development Bank (2001). *Istisna'a mode of financing*. wysiwyg://67/ http://www.isdb.org/english_docs/idb_home/MFIstMod_Home.htm [Accessed February 23, 2014].

Turner, DF (1979). *Quantity Surveying Practice and Administration*. London: George Godwin.

Chapter 8

The Application of Limited Liability in Construction Contracts from the Malaysian Law and Shariah Perspectives

ZUHAIRAH ARIFF Abdul Ghadas

Introduction

Malaysia's history was generally influenced by the Portuguese, Dutch and British (Aiman, 2000). Nevertheless, the English laws were the most influential in the Malaysian legal history. The first intervention in law-making and administration of justice by the British was initiated in the Straits Settlements (Hooker, 1969). This was possible because the British East India Company had great influence in the Straits Settlements. The company was given the authority to administer the law and as such the law applicable was English law. In 1878, the English Commercial law was formally introduced to the Straits Settlements by virtue of the Civil Law Ordinance. Similar to other common law principles, the practice of limited liability in Malaysia came along with the English influences.

From the legal perspective, limited liability is a term which has no precise definition. It is generally used to describe the situation where a person has done an act which, under the prevailing rules of the legal system, will incur a liability but is excused, wholly or partly from incurring that liability. Limited liability often arises either because the legal system provides the conveniences under which the parties may organize their affairs so as to achieve limited liability or because the legal system produces a rule which provides that, in specific circumstances, liability shall be limited. There are certain industries which may not exist without special provisions for limited liability, for example, the workers' compensation statutes generally state the limits of liability of employers and certain treaties such as the Warsaw Convention limit the damages for airline passengers.

The Historical Perspectives

The principal arguments in favor of limited liability stemmed from the common law rule of unlimited joint and several liabilities of partners in partnerships (Livermore, 1935). Under this principle, partners may be held liable for the whole amount of the firm's debts. The main argument against unlimited liability in partnership is that partners, particularly those who were not involved in committing the default, can be made liable for the loss suffered by the firm due to a default of one or some of the partners. By attaching the liability of the business to all partners as a collective of the firm, no partner can escape from the liability of the business, however innocent he or she is.

The rationale of joint and several liabilities is generally based upon the fundamental principle of agency law which is applied in partnership, whereby the firm, as the principal, will be liable for the act of its agents (the partners) who act within their authority. However, as the firm is a collective of all partners, it means that every partner will be liable for any act committed by their fellow partner in the due course of the partnership business. None of the partners is allowed to escape or limit their liabilities in the firm.

When companies were first introduced into the market, they also applied the unlimited liability concept as practiced in partnerships. However, due to the criticism towards the concept, limited liability was later made available to registered joint stock companies. Despite the contention that the concept of limited liability arises from the criticism of the joint and several liability rule practiced in partnerships, historical writings proved that the development of limited liability law is actually an ancient activity.

Under the European legal system, the municipalities and some private organizations of trade guilds or religious institutions were recognized as having corporate personality and separated from the people who made up the corpus (Carney, 2000). In the late medieval period, the Italians introduced a trading vehicle which involved an element of limited liability, known as the *compera*. Later, the Venetian and Genoan developed a better trading vehicle called the *commenda*. The developed form of *commenda* was later combined with the notion of transferability of shares and received into the continental legal systems as the *Societe en Commandite par Actions* and the *Kommanditgesellschaft auf Aktien*. Nevertheless, the origins and economics discussion of limited liability lay in England, whereby discussion was centered on the increasing agency cost of firms with passive investors and management separated from ownership rather than the efficiency of limited liability to externalize the costs of firm activities.

The early meaning of limited liability was merely confined to the separate legal personality of companies whereby the members could not be sued directly for the companies' debts. Even without the State-recognized corporate personality, many chartered corporations already practiced an express clause in their charter which provided that liability of shareholders was to be limited only up to the amount of their contribution to the joint stock. Common law by ad hoc also made successful attempts to contract out of limited liability by inserting a clause into all company's contracts that the other party could only take the company's assets for satisfaction of debts. In 1855, the Limited Liability Act was passed and limited liability became generally available in England. The Act granted limited liability to members of registered joint stock companies and this led to an increase in the number of companies seeking registration. While England was more industrialized than the United States during the 19th century, it lagged behind the latter in the development of limited liability. Forbes suggested that possible explanations for this was that the English entrepreneurs found limited liability less useful than the Americans and there was a lack of jurisdictional competition as regards the limited liability compared to those that existed in the United States.

Limited Liability in Contracts

The early practice of limited liability in contractual transactions can be traced back to Roman law (Hillman, 1997). It is clear that the gist of contractual obligations was actually personal, which meant that only the contractual parties were bound by the contract. The third parties could not be involved in the contract, even if they controlled one of the contractual parties. This practice was later recognized as an impediment to commercial transactions as it made conducting business through other parties difficult. The contract law was later developed to allow some interference from a third party. One sign of the change was the *actio institoria*, which allowed a claim to be made against the "principal" for acts done by the "agents" in the "course of the business" (Watson, 1965). Another important technique employed to limit personal liability in contracts is the *peculium*, whereby some assets owned by a slave owner or a father (*pater familias*) were entrusted to the slave or to the son, respectively and the commercially-minded slave or the son was encouraged by the master or father, to trade with the assets (*peculium*) (Johnston, 1995). However, debts and liabilities incurred in such trading could only be enforced by the third parties against the master or the father (*pater familias*) up to the extent of the *peculium*, and not against all of the latter's property. The *peculium* system allowed the Roman to invest in a business through his slave or son and limited his liability by fixing the size of the *peculium*. This business system had proved to be an excellent limited liability vehicle for conducting business activities and over time, the Romans adapted to the inevitable development of claims arising out of activities associated with the *peculium*.

The emergence of the *commenda* in Italy in the 11th century was another significant milestone in the history of limited liability in contracts. The *commenda* was found to be used largely in the sea trade (Postan and Habakkuk, 1966). It involved a system where one party would provide capital to another party to finance an overseas commercial venture. The capital provider would not be involved in both the voyage and running of the trade. He was some sort of a dormant partner. On the other hand, the other party, who accepted the capital, would have to endure the sea voyage and shouldered the responsibility of all aspects of management of the venture. The *commenda* would terminate upon completion of the venture whereby the parties would then divide the profits under a predetermined formula (Sapori, 1970).

The establishment of the *commenda* as an accepted form of investment enabled passive investors to enjoy limited liability. As such, they were able to diversify their investments by placing money in several *commenda* ventures rather than placing large bets on a single voyage (Postan and Habakkuk, 1966). The characteristics of *commenda* had been said to be the origin of the modern limited partnership, whereby one of the partners (limited partner) was allowed to limit the liability up to the amount of his contribution to the firm but he did not have any right in the management of the business. In 1408, Florence enacted a statute which allowed the creation of a limited partnership known as *Societe en Commandite*. The origin of the French limited partnership (*Societe en Commandite*) was traced to the time of Louis XIV in 1671 and later in the Code of Commerce of 1806. The *Societe en Commandite* which was largely used in France later became the model for the Irish Anonymous Partnership Act of the late 18th century.

In England, the development of limited liability in contractual transactions was found to be along a different line and at a slower pace than what occurred in its counterparts on the Continent (Perrott, 1982). Under common law, the concept of limited liability was introduced two hundred years ago in order to enable the large-scale investment necessary for the Industrial Revolution to take place (Crowther, 2000). With the severance of investment in the business from the management of that business there was considered a need for the protection of the investors, who were often individuals with a relatively small amount of capital, from the possible fraudulent actions of the managers of the business. This paved the way for the attraction of many more investors, thereby enabling the growth in size of the business enterprises, with those investors secure in the knowledge that they were protected from any loss greater than the sum they had invested in the enterprise. Thus for relatively small levels of risk they were able to expect potentially great rewards and thereby escape from some of the consequences of the actions of the enterprise.

In today's modern contracts, limited liability in contractual transactions is no longer an alien concept. It has been practiced through the incorporation of terms in the contractual document itself, for example, the limitation and exemption clauses or through direct application in the business vehicles. Companies with limited liability and limited partnerships are already a norm in the market place, whilst new expansions of the partnership structure such as the limited liability partnership has started to make its mark in the market place. The application of limited liability in contractual transactions, either in the past or today is basically the same whereby there is a limitation of liability for contractual parties as a consequence of breach of a contract. Creditors who are involved in the contract will normally try to seek damages to the maximum or ceiling and this is when limited liability is commonly applied. Where there is a right for the contractual parties to limit their liability, creditors cannot claim more than the amount stated in the contract even if they are entitled to more than what they are gaining. Nonetheless, being parties to the contract, these creditors, who are also known as voluntary creditors, may manage their risk through a range of contractual and other techniques, including diversification and insurance (Grantham and Rickett, 1998).

Limited Liability from Shariah Perspective

From the Shariah perspective, the approach on limited liability is quite distinct compared to the common law approach (Zuhairah Ariff and Engku Rabiah, 2009). The *sharikah* and *mudarabah* contracts have been utilized across the board, regardless of the formal business structures, i.e., whether they are partnerships or companies.

For example, if the parties opted for the partnership structure, under Islamic principles, the contract will still be that of *sharikah* or *mudarabah*. The liabilities of partners are still limited up to their capital contribution, unless they opt for *sharikah al mufawadah* where liabilities of partners are unlimited. Even if the parties choose the company structure, the contract will still be that of *sharikah* or *mudarabah*. The parties' liabilities remained the same under Islamic law, i.e., up to the amount of their capital contribution. There are no separate contracts on the basis of pure business structures — partnerships or companies. Thus, the paramount consideration in determining liability in Islamic law is not the business structure, but the actual *sharikah* contracts between the parties. If the parties want limited liability, they can choose *sharikah al 'inan* or *mudarabah*. If they want unlimited liability, they can choose *sharikah al mufawadah*. Thus, from the foregoing discussions we can see that the origin of limited liability regime in *sharikah* is the contract between the parties.

As regards to liabilities, *sharikah* generally refers to contractual obligations. It is observed that the discussion on liabilities of partners in *sharikah* in the event of liabilities exceeding the assets have not been elaborately made in the classical Islamic law literature (Zuhairah Ariff and Engku Rabiah, 2009). What has been mentioned is just the general principle that liabilities follow the amount of capital contribution. This lack of elaborate discussion is understandable because the way Islamic economics and business works ensures a built-in mechanism against excessive mismatch in asset and liability ratio. As pointed out by Umar Chapra (1985), in an Islamic economy, since all financial participation in business would be essentially in the form of equity, the only exceptions being suppliers' credits and *qurud hasanah* (beneficial loans), the liability of the partners would in reality be limited to their capital contributions. Prudence would induce the suppliers to keep an eye on total equity, movement of sales and cash flows of the business concerned, while *qurud hasanah* would tend to be limited. All other participants in the business (whether by way of loan or equity) would be treated as equity holders and would share in the risks of business. Since interest bearing loans are not allowed, the total obligations of the business could not be out-of-step with the total assets, and any erosion in their value may not exceed the total equity. Hence, in the ultimate analysis liability would essentially be limited to the extent of the total capital (including ploughed-back profits) invested in the partnership business.

There are many ways of categorizing *sharikah* (Zuhairah Ariff and Engku Rabiah, 2009). The classical categorization of *sharikah* is based on a variety of factors. If origin of the partnership becomes the determining factor, *sharikah* can be divided into two broad categories, namely, *sharikah al mulk* (proprietary partnership) and *sharikah al 'aqd* (contractual partnership). For *sharikah al mulk* (proprietary partnership), the origin of the partnership is the joint ownership of property. Joint ownership is its only qualification, and no joint exploitation of property is necessary. It occurs when two or more people are partners in the possession of property. The rule governing this type of *sharikah* is that any increase in the property shall be shared by the co-owners in proportion with the extent of their ownership. Each of them is in the category of a stranger in regard to any action on the part owned by his colleague. In other words, it is not lawful for either partner to perform any act with respect to the other's share except with the latter's express permission (Al Qusi, Abdul Mun'irn, 1982). Thus, in terms of liability of the partners, they are quite independent of each other, except for actions based on express authorization by any of the partners. Their partnership is only in terms of ownership and potential sharing of any profit or increase in the co-owned property, not in terms of sharing the liabilities arising from the partners' actions. This type of *sharikah* may not be known in the common law or Malaysian law. In fact mere joint-ownership is generally insufficient to constitute a partnership in common and Malaysian law.

For *sharikah al 'aqd* (contractual partnership), the origin of the partnership is the contract between the parties. The structure of this type of *sharikah* may have more similarities with the normal partnership in common law and Malaysian law. For *sharikah al 'aqd*, joint ownership is not an element necessary for the establishment of the partnership. The emphasis is rather on the joint exploitation of capital and the joint participation in profits and losses, based on the terms of the partnership contract. Joint ownership is one possible consequence, and not a prerequisite for the formation of *sharikah al 'aqd* (Al Qusi, Abdul Mun'irn, 1982).

The jurists further sub-divide *sharikah al 'aqd* into various other categories. The sub-divisions depend on a number of factors. If the underlying factor is the subject matter of capital contribution, *sharikah al 'aqd* can be sub-divided into three main categories, namely, *sharikah al amwal, sharikah al a'mal* and *sharikah al wujuh*. When the subject matter of the capital is money, it becomes *sharikah al amwal* (monetary partnership). If the capital is in the form of labor, it becomes *sharikah al a'mal* (labor partnership). If the capital is in the form of reputation or creditworthiness, it becomes *sharikah al wujuh* (reputation partnership).

The jurists also make further sub-divisions to *sharikah al 'aqd* based on the terms of the contract, i.e., whether the partners are required to contribute equally to the capital and enjoy full equality in exploiting the capital and sharing the profit. Based on this consideration, *sharikah* can be divided into two types, *sharikah al mufawadah* and *sharikah al 'inan*.

Basically, *sharikah al mufawadah* means an unlimited investment partnership, whereby each partner must contribute equally to the capital, and enjoys full and equal authority to transact with the partnership capital or property. The Hanafis consider each partner as an agent (*wakil*) for the partnership business and stands as surety (*kafil*) for the other partners. Thus, the partners can be made jointly and severally responsible for the liabilities of their partnership business provided that such liabilities have been incurred in the ordinary course of business (Umar Chapra, 1985). This type of *sharikah* clearly implies unlimited liability on the part of partners since they are both agents and guarantors of each other.

On the other hand, *sharikah al'inan* can be loosely defined as a limited investment partnership. Whereby each partner may only transact with the partnership capital according to the terms of the partnership agreement and to the extent of the joint capital. Hence, their liability towards third parties is several but not joint. In other words, the liability of partners in *sharikah al'inan* resembles that of modern-day limited liability partnerships.

Both *sharikah al-mufawadah* and *sharikah al'inan* can occur in all the three earlier types of *sharikah*, i.e., *sharikah al amwal* (monetary partnership), *sharikah al a'mal* (labor partnership) and *sharikah al wujuh* (reputation partnership).

The jurists differ with regards to a special type of commercial dealing, i.e. *mudarabah* (profit-sharing) (Zuhairah Ariff and Engku Rabiah, 2011);[1] whether it is a kind of *sharikah* or not. Some of them, i.e. the Malikis and Hanbalis regard it as a form of *sharikah*, while others, i.e. the Hanafis and Shafi'is categorize it as a separate kind.

Mudarabah is basically a form of commercial arrangement where one of the contracting parties act as the provider of capital (termed as *rabb al mal*), while the other party acts as the entrepreneur (termed as *mudharib*). The essential difference between *mudarabah* and other forms of *sharikah* is whether or not all the partners make a contribution towards the capital as well as management of the partnership, or only one of these. In *mudarabah*, one party provides capital whilst the other provides management skill. In *sharikah*, all partners contribute to both capital and management of the partnership.

In *mudarabah*, the *rabb al mal* is the dormant partner, while the *mudharib* is the active partner who provides the entrepreneurship and management for carrying any venture, trade or industry with the objective of generating profit. Any accruing profit shall be shared between the *rabb al mal* and *mudharib* according to a pre-fixed ratio. In the event of loss, the *rabb al mal* bears the financial losses to the extent of his contribution to the capital, while the *mudharib* suffers the frustration of a fruitless effort. Again, in the *mudarabah* arrangement a limited liability regime is created. However, the regime is quite different from modern limited liability. On the one

---

[1] Also known as *qirad* or *muqaradah*. It has been said to be the possible origin for the *commenda* of medieval Europe, see for example, Hillman, RW (1997). Limited liability in historical perspective. *Washington and Lee Law Review*, 54, 621–622.

hand, in *mudarabah*, it is the active partner who is exempted from financial liability (except if proven negligent or fraudulent). On the other hand, though the passive partner bears the bulk of financial liability he also enjoys limited liability because his financial liability is just to the extent of his capital contribution.

From the many types of *sharikah*, the one mainly used in contemporary Islamic banking and finance is that of *sharika al 'inan* in the category of *sharikah al amwal*. *Sharikah al mufawadah* is rarely opted for due to the higher degree of responsibility and the practical difficulty to achieve full equality between the partners in all aspects of the partnership. Another commonly utilized contract is that of *mudarabah*, which, some jurists consider distinct from the other forms of *sharikah*. Actually, *mudarabah* can be construed as a *sharikah* with monetary capital on the one part and labor on the other.

It is observed that despite the different approaches between the common law and Shariah, it is apparent that the principle of limited liability is viable and acceptable under the Shariah.

Limited Liability in Construction Contracts

As stated earlier in this paper, the doctrine of limited liability is used to excuse, wholly or partly from incurring liability by organizing terms of the agreement so as to achieve limited liability. In PAM 2006 and PWD 203A (Rev. 1/2010) standard forms of contract, the application of limited liabilities by the employer or government and the contractors are obvious throughout the contract. Examples of terms which relate to limitation of liabilities are clauses on contractors' obligations, defects liability, design liability, insurance requirement, retention fund and performance bond. For the purpose of discussion, this chapter will only discuss on these terms although there are other terms which are directly and indirectly related to limited liability in Malaysia's standard forms of construction contracts.

Obligation Clauses

Under both PAM 2006 and PWD 203A (Rev. 1/2010), emphasis is made on the compliance of the employer's or government's and contractors' respective obligations under the contract.[2] This means both the contractor and employer or government cannot be made liable for any other obligations outside the contract. For example, if additional works or materials which are not specified in the contract are required in order to complete the work, the contractors are not obliged to do the additional works or supply the materials unless such additions are requested formally and approved by the employer or government under the variation clause.[3] Vice-versa, the employer is not obliged to pay the contractors for any additional works or materials which are not specified in the contract unless such matters are approved by them as variation.[4] However, the scope of liability of contractors under PWD 203A are wider compared to obligations of contractors under PAM 2006 as under PWD 203A, other than the obligations specified as terms of the contract, the contractors have to carry out any other obligations and responsibilities under the contract.[5]

The employers are also limiting their liability to pay the contractor only up to the amount of the contract and not on the total amount incurred by the contractor which may exceed the contract price even though such costs were incurred to complete the works.[6]

Design Liability

As regards to liability in design, there is a clear limitation of liability enjoyed by the contractors if they are not involved in the design as the PWD 203A and PAM 2006 clearly stated that where the contractors are not involved in designing, they will not be liable for the defect in design.[7] In *Moneypenny v. Hartland* [1824] 1 Car. & P. 351 the Court held that in the absence of proper delegation, the designer is not entitled to rely upon the work of other. In fact, in *University of Glasgow v. William Whitfield and John Laing Construction*

---

[2]Clause 1.0 of PAM 2006; Clause 10.0 of PWD 203A (Rev. 1/2010).
[3]Clause 11 of PAM 2006; Clause 24.0 of PWD 203A (Rev. 1/2010).
[4]*Ibid.*
[5]Clause 10(k) of PWD 203A (Rev. 1/2010).
[6]Clause 13.0 of PAM 2006; Clause 7.0 of PWD 203A (Rev. 1/2010).
[7]Clause 1.2 of PAM 2006; Clause 22.1(a).

[1988] 42 BLR 66, the Court held that contractors have no duty to warn the client of design deficiencies.

Defects Liability

Different from design liability, the application of limited liability for defects liability is not available for contractors but enjoyed by the employer or government.[8] In *P & M Kaye Ltd v. Hosier & Dickinson Ltd* [1972] 1 WLR 146. Lord Diplock held that the contractor has the liability to mitigate the damage caused by his breach by making good defects of construction at his own expense. Failure of the contractors to remedy the defect shall entitle the employer or government to employ another contractor to remedy the defects and deduct the cost from the payment due to the contractor or recover from the retention fund or performance bond.[9]

Insurance Requirement

One of the main mechanisms to limit liability of both parties under PAM 2006 and PWD 203A (Rev. 1/2010) is the insurance requirements. There are terms on insurance-imposed obligations to the contractors to take insurance coverage against personal injuries, damage to property and against loss and damage to works.[10] Such requirements allow the contractors to enjoy some limitation of liability by transferring the liabilities to the insurance companies.

Different from PWD 203A contract which does not impose insurance requirements on the government, PAM 2006 contract requires the employer to take up insurance for new buildings works and insurance for existing buildings or extension of buildings.[11]

---

[8]Clause 15.3 and 15.4 of PAM 2006; Clause 48.1 (a), (b) PWD 203A (Rev. 1/ 2010).
[9]Clause 48.2 of PWD 203A (Rev. 1/2010); Clause 15.5 of PAM 2006.
[10]Clause 18, 19, 20.A, 20.B, 20.C of PAM 2006; Clause 15 and 18 of PWD 203A (Rev. 1/2010).
[11]Clause 20.B and 20.C.

Retention Fund and Performance Bond

The requirements for payment of performance bond and retention fund in PAM 2006 and PWD 203A are also seen as a mechanism to limit liability of both parties. The retention fund and performance bond allows the contractors to limit their liabilities for defects during the Defects Liability Period (DLP) up to the amount of the retention fund or performance bond. The retention fund and performance bond also serve as a limited liability mechanism to the employer by ensuring that any cost to remedy defects during the DLP shall be imposed on the contractor and not the employer or government.

Conclusion

The application of limited liability in contract is applicable and acceptable under both Civil law and Shariah but the way they are construed are a bit different, whereby limited liability in Civil law may be derived not only through business arrangements or terms of contract but also via the business vehicle or entity whilst under Shariah, limited liability is only available via agreements of the parties and not through the business structure or entity.

As for the construction contracts, it is obvious that the application of limited liability via terms of the contract are extensively applied. Such application is seen vital due to the nature of construction business which is high risk and full of unforeseeable and uncalculated risk.

List of Cases

Moneypenny v. Hartland [1824] 1 Car. & P. 351
University of Glasgow v. William Whitfield and John Laing Construction [1988] 42 BLR 66
P & M Kaye Ltd v. Hosier & Dickinson Ltd [1972] 1 WLR 146

References

Aiman, N (2000). Public regulation of companies: The investigatory powers of the corporate regulators — An overview of Malaysian law with comparisons with

Australia and United Kingdom, unpublished Ph.D. Thesis, Bond University.

Al Qusi and Abdul Mun'irn (1982). *Riba*, Islamic law and interest. Unpublished Ph.D. Dissertation, Temple University.

Carney, WJ (2000). *Limited Liability, Encyclopedia of Law and Economics*, University Utrecht. https://reference.findlaw.com/lawandeconomics/5620-limited-liability.pdf. [Accessed on November 1, 2018].

Government of Malaysia (2010). Standard Form of Contract to be Used Where Bills of Quantities Form Part of the Contract. Public Works Department (PWD) Form 203A (Rev. 1/2010).

Grantham, R and C Rickett (1998). *Corporate Personality in the 20th Century*. Oxford: Hart Publishing.

Hillman, RW (1997). Limited liability in historical perspective, *Washington and Lee Law Review*, 54, 615.

Hooker, MB (1969). East India Company and the Crown, *Malaya Law Review* 1, 11(2), 1773–1858.

Johnston, D (1995). Limiting liability: Roman law and the civil law tradition, *Chicago-Kent Law Review*, 70, 1521–1523.

Livermore, S. (1935). Unlimited liability in early American corporation, *Journal of Political Economy*, 43(5), 674–687.

Malaysia Institute of Architects (Pertubuhan Arkitek Malaysia — PAM) (2006). Agreement and Conditions of PAM Contract 2006 (With Quantities).

Perrott, DL (1982). Changes in attitude to limited liability — The European experience. In *Limited Liability and The Corporation*, Tony Orhnial (ed.), Sydney: Law Book Co. of Australasia.

Postan, MM and HJ Habakkuk (eds.) (2000). *The Cambridge Economic History of Europe*. Cambridge: Cambridge University Press.

Sapori, A (1970). *The Italian Merchant in The Middle Ages*. England: Norton Publishing. Crowther, D (2000). Limited Liability = Limited Risk = Limited Accountability. http://www.le.ac.uk/ulmc/research/crowther.pdf

Umar Chapra (1985). *Towards A Just Innately System*. Bradford-on-Avon, Wiltshire: The Islamic Foundation Bradford Dotesios (Printers) Ltd.

Watson, A (1965). *The Law of Obligations in the Later Roman Republic*, pp. 191–192. Oxford: Clarendon Press.

Zuhairah Ariff, AG and A Engku Rabiah (2009). Partners limited, limited liability in partnerships structure: An overview of the English law and the Islamic law, Shariah Law Reports, No. 1, *LexisNexis Malayan Law Journal*.

Zuhairah Ariff, AG and A Engku Rabiah (2011). The development of partnership based structure in comparison to the concept of Musharakah (Sharikah) with special reference to Malaysia. *Journal of Islam in Asia*, Special Issue, 2, 293–315.

Chapter 9

The Application of Shariah Principles of ADR in the Malaysian Construction Industry

ZUHAIRAH ARIFF Abdul Ghadas, ROZINA Mohd
Zafian and ABDUL MAJID TAHIR Mohamed

Construction Disputes

The construction industry is known for its conflict, with its characteristic mix of complex contractual relationships, huge sums of money at stake, highly complex projects and remorseless time pressure, as much as its spectacular construction and civil engineering projects (Holtham *et al.*, 1999; Mackie *et al.*, 2000). It also has a reputation as a tough and aggressive world in which the weakest and even at times some of the strongest will be driven up the wall (Mackie *et al.*, 2000). Disputes result not only from destructive or unhealthy conflict, but also when claims are not amicably settled (Mohan, 1998). Hence, a construction project is considered by many a dispute waiting to happen (Patterson and Grant, 2001).

Construction disputes typically comprise both technical and legal dimensions (Cheung, 2006), the former being the dominant issue in disputes. For this reason, litigation may not be the most appropriate forum for dealing with these types of disputes. The dissatisfaction with the traditional dispute resolution mechanisms which can no longer successfully cope with the growing needs and challenges of the present construction environment has invoked the industry to look towards other alternative methods (Pēna-Mora *et al.*, 2003). Alternative Dispute Resolution (ADR) is a generic description used to identify a wide range of resolution processes that aim to resolve disputes speedily and cost-efficiently (Cheung, 2006).

Disputes within the construction industry are inevitably related to time, money and quality. Disputes that are not resolved promptly, in all probability, would incur a considerable escalation in expenses which are hard or impossible to quantify. The visible expenses anticipated include the legal representatives, expert witnesses and the cost of the dispute resolution proceedings itself. Amongst the less visible costs would be the company resources assigned to the dispute and lost business opportunities, while the intangible costs are identified as detriment to good working relationships and potential value lost due to inefficient dispute resolution process.

Alternative Dispute Resolution (ADR)

Over the last few decades the perceived shortcomings of litigation and also arbitration have resulted in attempts to find other quick means to resolve construction disputes (Treacy, 1995). Alternative Dispute Resolution was first developed in the United States in early 1980s as a result of dissatisfaction with the delays, costs and inadequacies of the litigation process (Mackie *et al.*, 2000). However, it only began to receive consideration in the late 1980s and early 1990s. Since then, its development as a process to resolve civil disputes relatively inexpensively and quickly has gained momentum and is now widely practiced in construction industries in many countries, especially in Canada, United Kingdom and Australia. The acronym ADR has also been defined as Additional Dispute Resolution and Assisted Dispute Resolution. With time, it also stands for Appropriate Dispute Resolution (Mackie *et al.*, 2000)[1] and Amicable Dispute

---

[1]Mackie *et al.* (2000) indicate that most advocates of ADR agree that the term "alternative" is inappropriate as it adds to, and not replaces the litigation option.

Resolution to reflect these desired outcomes efficiently (Cheung, 2006). The realization of ADR as a process that complements both litigation and arbitration has meant that the processes are constantly expanding to include new techniques which offer no limits to the types of dispute resolution processes that can be utilized. The main attraction of ADR is often the consensual process, but this also means that it will not be successful unless the parties each have a genuine desire to reach a settlement. Even though the most common ADR methods do not provide assurance of a resolution, in practice most of these methods lead to a final settlement. The key to a settlement process is that the parties and those assisting in the process understand and agree to the same process.

The reasons for resorting to ADR include time savings, less costly discovery, more effective case management, confidentiality, and facilitation of early, direct communication and understanding among the parties of the essential issues on each side of the dispute (Treacy, 1995). Other reasons are preservation of ongoing party relations, savings in trial expenses and providing qualified, neutral experts to hear complex matters. Traditionally, arbitration was the forum sought in the construction industry (Battersby, 2001). The ADR processes differ in their formality and placement of decision-making power. Some methods are non-binding and allow the parties to have control at all times over the outcome of the dispute, participate in the development of an agreeable settlement in the presence of a neutral third party and withdraw from the process at any point (Pēna-Mora *et al.*, 2003). Other methods may become binding where all power lies with the neutral third party which is mandatory and have a formal structure that require strict adherence to the rules and implementation (Uff, 2005). The process chosen should provide a solution to suit the varied nature and size of construction disputes with the object of saving time and costs (Battersby, 2001). Furthermore, due to the divergence in construction disputes, the right process should also be adapted to the type of problem (Mackie *et al.*, 2000).

Apart from arbitration, other ADR methods include mediation, conciliation, early neutral evaluation, expert determination and mini-trial as well as other hybrid methods such as med-arbitration and dispute adjudication or review board. Brown and Marriot (1999), have identified eighteen main dispute resolution methods ranging from processes which offer the least control, which is litigation, to those that offer the greatest control, that is, negotiation. Due to the divergence in construction disputes, Mackie *et al.* (2000) is of the view that the right ADR process should be adapted to the type of problem.

Although there is no one exclusive ADR for the construction industry, apart from arbitration, which is the most widely-used form of ADR mechanism in the construction industry, other spectrums of ADR include negotiation, mediation, conciliation, med-arbitration, adjudication, mini-trial, expert determination or appraisal, court-annexed ADR and dispute review board (Brown and Marriot, 1999). The array of methods has advantages and disadvantages and despite having similar objectives, the processes involved are significantly different from one method to the other (Battersby, 2001).

ADR in Malaysia

Notwithstanding this wide adoption of ADR within the construction industry, the geographical differences attributed to cultural factors, maturity of the industry and prevalent legal systems in force influences the use of ADR practices (Cheung, 2006). Furthermore, participation in ADR techniques remains largely voluntary, and the legal implication arising from them remain uncharted.

The most common ADR methods to resolve disputes in the construction industry are arbitration, mediation or conciliation, adjudication and expert determination (Battersby, 2001). However, the dispute resolution methods that are normally incorporated in Malaysian construction contracts are arbitration and mediation (Lim and Xavier, 2002). Arbitration is recognized and practiced worldwide while mediation is yet to gain the popularity that arbitration has achieved in the resolution of commercial disputes in Malaysia (Xavier, 2002). At present, Malaysia is seeking for an efficient and economical dispensation of justice and a more suitable dispute resolution technique to deal with current and future challenges in the construction industry.

This chapter discuss the three of the aforementioned ADR methods, namely arbitration, mediation and adjudication.

Arbitration

While the court is the main forum for resolution of construction disputes, arbitration is a well-established part of the Malaysian construction industry. Arbitration has been in use in the region of Asia for quite some time and its provisions are included in almost all construction contracts in this region (Battersby, 2001). Malaysia is not precluded and arbitration clauses are found in standard forms of contract, which are the PWD 203 and 203A (Rev. 1/2010)[2] series which are used in public sector works, and the PAM 2006[3] and CIDB Building Works 2000 Edition[4] which are used in private sector works. All these standard forms provide for arbitration as the final form of dispute resolution and has produced a *de facto* universality of arbitration as the normal method of settling disputes (Sundra Rajoo and Davidson, 2007). It is a method of private dispute resolution in which the parties to the dispute agree to have it settled by an independent third party and to be bound by the decision he makes. This agreement may also be entered into after the dispute has arisen. The arbitrator may be chosen by agreement between the parties themselves or may be appointed by a nominating body named in the contract.

Amongst the advantages of arbitration is the privacy and confidentiality afforded to the parties. The parties also have the freedom to determine an arbitrator or appropriate appointing body to ensure that he or she has relevant expertise and experience. They are free to choose their own rules, with great procedural and substantive flexibility. There are very limited grounds of appeal against an

---

[2]Clause 66 of PWD 203A (Rev. 1/2010).
[3]Clause 34 of PAM 2006.
[4]Clause 47 of CIDB Standard Form of Contract for Building Works 2000 Edition.

arbitration award.[5] One of the disadvantages of arbitration is that the flexibility of the process can create uncertainty among the parties. Depending on the circumstances of a dispute, arbitration can be very quick and cost-effective as a means of resolving dispute. On the other hand, it can also be very time-consuming, cumbersome, expensive and adversarial which contributed to it earning the name "litigation in the private sector."[6]

In Malaysia, the guiding principles are set out in the Arbitration Act 2005, which repeals and replaces the Arbitration Act 1952. It is applicable to all arbitration proceedings commencing after March 15, 2006. This new Act also addresses some of the perceived and actual failures of the arbitration process in the previous Act. It is based on the United Nations Commission International Trade Law (UNCITRAL) Model Law on International Commercial Arbitration ("the Model Law") and has been adopted by some 63 countries worldwide (Sundra Rajoo and Davidson, 2007). Although the Model Law was primarily drafted for the purpose of international arbitrations, the member states are free to modify the Model Law for use in their domestic arbitral regime. Among the common law states that have adopted this suggestion is India and New Zealand. Notwithstanding certain distinctions between their international and domestic arbitration, the New Zealand Arbitration Act applies the provisions of the Model Law in both these regimes. In relation to this, the Malaysian Arbitration Act closely resembles the New Zealand model.

Although arbitration is a consensual process, the jurisdiction of the arbitrator or the arbitral tribunal and the scope of the arbitration are fixed by the terms of the arbitration agreement and the arbitration will be conducted according to certain prescribed procedural rules. These rules may be expressly agreed by the parties but where the parties do not make such a choice, or where the rules

---

[5]Sections 15(5) and 18(10) of the Malaysian Arbitration Act 2005.
[6]See the judgment in *Northern Regional Health Authority v Derek Crouch Construction Company Limited* [1984] 1 QB 644 at 70 where Sir John Donaldson MR stated that "Arbitration is usually no more and no less than litigation in the private sector." See also Battersby.

which they choose are silent on a particular point, the procedure of the arbitration will be governed by the statute on arbitration of the country in which the proceedings take place. In Malaysia, the rules for arbitration which the parties may agree to submit to are the PAM Arbitration Rules, The Institution of Engineers Malaysia (IEM), Rules for Arbitration of the KLRCA,[7] ICC Rules of Arbitration or Malaysian Institute of Arbitrators (MIArb) Arbitration Rules 2000 Edition.[8] By virtue of the doctrine of severability provided in the Act, the arbitration clause agreement has a separate existence and is an agreement independent of the main contract.[9]

Arbitration is to be distinguished from other ADR methods, in that it is a judicial process involving evidence and submissions by the parties before an independent arbitrator, who subsequently reaches a decision on the parties' respective rights and obligations under the contract. The decision of the arbitrator is final and binding, subject to a possibility of very limited intervention by the courts. In contrast with adjudication, arbitration is not confined to certain time limits, more readily enforceable and not susceptible to

---

[7]The Kuala Lumpur Regional Centre for Arbitration (KLRCA) was established in 1978 (under the auspices of the international governmental international law body, the Asian-African Legal Consultative Organization (AALCO) to provide a forum for the settlement of disputes by arbitration and also mediation in trade, commerce and investment within the Asia-Pacific region. While it has the support of the Malaysian government, KLRCA is a non-profit organization and is not a branch or agency of the government. In relation to arbitration, the KLRCA adopts the UNCITRAL Arbitration Rules of 1976 with certain modifications and adaptations. http://www.rcakl.org.my/ [Accessed January 7, 2010].

[8]MIArb Arbitration Rules 2000 Edition are up-to-date and aim to overcome the common pitfalls of arbitration as well as ensuring fair and speedy disposal of disputes. The rationale of MIArb Arbitration rules include recognition of party autonomy; flexibility for modification or simplification of arbitration procedure to suit nature and extent of disputes; facilitating expeditions proceedings; deferment of case stated or challenge of interim awards or decisions until after the main award or substantive merits are made, yet preserving a party's right to challenge an interlocutory decision or interim award. The MIArb Rules can be adapted for use in complex or simple arbitrations in different fields or industries.

[9]Section 18(2) of the Arbitration Act 2005. See also Sundra Rajoo and Davidson (2007) at p. 88.

a rehearing. In comparison to mediation, the parties may be compelled to submit to arbitration. A majority of construction contracts incorporated arbitration clauses as it is regarded as the main dispute resolution process, which is able to achieve the objectives of ADR. Yet, parties to construction disputes are often dissatisfied with the outcome of arbitration. Some of the discontentment spring from the following reasons.

Construction arbitration can last for many months or even years due to massive documentation to cover and consider use of experts, representations from various parties and the attempt to be as thorough as possible. Arbitration can be more expensive if more than one arbitrator is appointed in the tribunal. The efficiency of a process may be affected if a non-construction arbitrator is appointed when an arbitrator with technical background would be more appropriate for a particular dispute. Where disputes involve a myriad of technical as well as legal issues, arbitrators are not totally equipped to deal effectively with all the issues and the process reverts back to one that relies far more on adversarial strengths (Premaraj, 2007).

There is no recourse if an arbitrator makes a wrong decision. Protracted arbitration proceedings not only increase the costs of arbitration but are likely to cause immense harm to business relationships. As the construction industry normally involves complex processes and multi-disciplinary inputs, this may give rise to complex disputes, which can only be brought to arbitration upon completion of the project. Small disputes which are not compounded as early as possible may lead to large disputes which are more difficult to resolve. In the meantime, this chain of events would have starved the aggrieved party of cash flow which is vital to the completion of the project.

Mediation

A fundamental of mediation that involves the encouragement of settlement by the assistance of a third party has been a practice of the Eastern region for centuries. (Abraham, 2006; Bagshaw, 2008). The roots can be traced back to the Islamic *sulh* which covers negotiations, mediation, conciliation and compromise of action, the Chinese *xieshang* which means negotiation and consultation, and the Hindu's *panchayat* which represents a village tribunal of five elders.

Although the modern or formal mediation is not yet a common method of the dispute resolution process in Malaysia (Lim, 2004), the promotion of mediation in a number of industries have demonstrated that mediation is increasingly advancing into the society. The insurance and financial industries have established a single forum, known as the Financial Mediation Bureau (FMB), an integrated dispute resolution center for financial institutions under the supervision of the Central Bank of Malaysia.[10] This bureau that was set up on the basis of the ombudsman schemes in the UK offers consumer protection with regard to fair dealing with policy holders subject to certain requirements and limitations. Its main aim is to provide services with regards to complaints or disputes with financial service providers over a claim involving monetary loss arising out of services provided by the bank, finance or insurance company.[11] The award or decision of the FMB is binding on the institutions but not the complainant.

Other industries which provide an alternative forum facilitated by statute for resolving disputes or claims, which are simple, inexpensive and fast, are the Tribunal for Consumer Claims, an independent body operating under the Ministry of Domestic Trade, Co-operatives and Consumerism Malaysia, and the Tribunal for

---

[10]Prior to 2005, there were two separate bureaus: Insurance Mediation Bureau and the Banking Mediation Bureau. On January 20, 2005, The Central Bank of Malaysia officially launched the Financial Mediation Bureau (FMB) which replaced the IMB. The FMB combines the avenues for redress of the insurance industry and the banking and other financial services industry into a single organization. It was acknowledged at the launch that the two existing bureaus i.e. the IMB and the Banking Mediation Bureau had achieved considerable success in mediating disputes between insurance companies and banks and their customers. Rozina Mohd Zafian (2013). Legal analysis of statutory adjudication in the construction industry: special reference to UK and Malaysian acts. Unpublished PhD Thesis. International Islamic University Malaysia.

[11]Financial Mediation Bureau. http://www.fmb.org.my [December 23, 2009].

Homebuyer Claim operating under the Ministry of Housing and Local Government Malaysia. However, these bureaus and tribunals which are in operation are not strictly a mediatory forum as the mediators will evaluate the dispute or claim and make a final and binding award in the circumstances the parties fail to reach and agree to a settlement (Noor Azian, 2003).

In contrast with the abovementioned industries, there is no investigative complaints bureau in respect of the construction industry in Malaysia. The Malaysian CIDB is the only body at present that is attempting to regulate the construction industry and to promote standards within the industry. However, their role does not extend to investigating complaints from the public and other members of the industry and is simply an arm of the government, empowered to carry out its activities by the Minister of Works and is therefore not an independent body.

The concept of mediation is a totally different process from arbitration in all respects save only for the parties' agreement to utilize the process as an alternative to litigation and the objective of privacy (Battersby, 2001). It is non-binding and involves a neutral third party that does not make decisions. In construction mediation, this often forms part of the process of mediation in appropriate circumstances. It is contended that mediation and adjudication is included in a contract not as a replacement for arbitration but only as a means of avoiding arbitration. Mediation is faster and more cost-effective than arbitration. It also avoids the risk of win-loss situation. The parties to mediation retain control over their positions and can walk away from mediation or take time to reconsider the situation. When goodwill exists between the parties, mediation being non-adversarial helps to promote amicable settlements and preserves business relationships.

Although mediation of construction disputes highly resembles other mediation, there are some peculiarities which merit consideration. Normally a mediator chosen to deal with a construction dispute is likely to possess substantial knowledge and experience in the construction industry, thus saving time and expenses for the parties. As construction disputes are document-sensitive, the mediator will most probably be called upon to facilitate the amicable exchange of documents. The mediator may require a longer time for presentations from parties and caucuses with parties as it may involve multiple parties and complex issues. The mediator may also be required to render advisory opinion on matters if this approach is agreed by the parties.

In Malaysia, mediation is gaining recognition in the construction industry, which is evidenced by the incorporation of mediation terms as a first tier of dispute resolution in a number of the Malaysian construction contracts. In the CIDB Standard Form of Contract for Building Works 2000 Edition[12] and the PAM 2006[13] parties are encouraged to attempt to settle their disputes amicably by mediation prior to referral to other dispute resolution prescribed in the contract. In relation to these, there are many choices of rules that have been published by different bodies suitable for the Malaysian construction industry. Amongst the rules on arbitration are the PAM Mediation Rules and the CIDB Mediation Rules which are to be used in conjunction with their respective forms of contract, while the Rules for Conciliation/Mediation of the KLRCA, Chartered Institute of Arbitrators (CIArb) Mediation Rules, the Malaysian Mediation Centre Mediation Rules[14] or the ICC ADR Rules[15] are stand-alone rules that may be agreed upon by the parties.

It is hoped that the much-awaited judicial reforms in Malaysia which includes the setting up of a mediation system under a Mediation Act would not only require parties to mediate prior to

---

[12]Clauses 47.2 and 47.3 of CIDB Standard Form of Building Contract 2000 Edition.
[13]Clause 35 of PAM 2006.
[14]The Malaysian Mediation Centre is under the auspices of the Malaysian Bar Council.
[15]The International Chamber of Commerce (ICC) sets out these amicable dispute resolution rules, titled the ICC ADR Rules (the "Rules"), which permit the parties to agree upon whatever settlement technique they believe to be appropriate to help them settle their dispute. In the absence of an agreement of the parties on a settlement technique, mediation shall be the settlement technique used under the Rules. Retrieved from International Chamber of Commerce. http://www.iccwbo.org/uploadedFiles/Court/Arbitration/other/adr_rules.pdf [Accessed January 7, 2010].

filing in court, but also assist in clearing the backlog of civil cases. It was agreed that a court-mandated mediation system should be set up as mediation did not work well if it is outside the court system.[16]

Adjudication

Adjudication is a term long known outside the construction industry, with many and various meanings (Gaitskell, 2007). Within the construction industry, the advent of main contractors outsourcing their work has created problems, particularly in making payments to those down the construction chain. Since arbitration proved ineffective for such a dispute (Hibberd and Newman, 1999), adjudication made its first appearance in the United Kingdom construction industry in 1976 through its inclusion in the Joint Contracts Tribunal (JCT) sub-contracts form. Although adjudication was known then, it was seldom utilized due to its limited scope for disputes related to set-offs only. In the mid-1980s, adjudication was employed in some bespoke contracts mainly involving large projects in the UK (Stevenson and Chapman, 1999). This concept was later expanded in various forms to cover a full range of construction projects and disputes (Groton *et al.*, 2001).[17]

Adjudication is regarded as the nearest process to arbitration.[18] The principal advantage of adjudication over arbitration is that it is quick and relatively cheap. In contrast with mediation, adjudication results in a decision which is temporarily binding until finally determined by litigation, arbitration or settlement agreement between the parties.

---

[16]A statement by Khutbul Zaman, Bar Council Alternative Dispute Resolution Committee Chairman in Mediation system to tackle cases fast taking shape, *New Straits Times*, July 3, 2009.

[17]Groton, JP *et al.* (2001) provide that the English Channel Tunnel project as an example used a designated cadre of neutrals consisting of a three-member panel. This panel would be assigned on an *ad hoc* basis to deal with disputes as theory arose; the decisions of these panels were binding on the parties for the duration of the construction project, but were subject to appeal in arbitration.

[18]Opinions of HHJ Lloyd QC in *Glencot Development v Ben Barrett* [2001] BLR 207, at paragraph 19 and in *Balfour Beatty v London Borough of Lambeth* [2002] BLR 288, at paragraph 29.

Whilst there are obvious advantages to the industry in making contractual provision for binding interim decisions by an independent third party, one of the reasons for the lack of progress in this dispute resolution process is the common misconception that the engineer, architect or superintending officer (S.O), by virtue of his or her detailed knowledge of the contract, on behalf of the employer is already carrying out an adjudicatory function and should best be able to find common ground for settlement (Sykes, 1996). The preliminary reference to the architect, engineer or S.O is usually the first tier to resolve disputes unless and until the aggrieved party refers the matter to arbitration, to commence only on completion of the works. Whilst this position is less than satisfactory, given that the engineer, architect or S.O himself or herself may have been the cause of the dispute, or is not regarded as an impartial and informed third party, the disputing party often prefers to refer any dispute on such decisions straight to arbitration or litigation rather than to make use of a separate and independent expert adjudicator. Compared to arbitration, which the courts are generally respectful of a parties' decision to arbitrate, without an underpinning legislation to encourage adjudication as an interim binding decision, it is more likely than not that the courts would hold an adjudicator's decision as having "an ephemeral and subordinate character" and would therefore not treat such a decision on the same footing as an award made under an arbitration agreement despite its binding nature under the contract between the parties.

In summary, adjudication can be described as a procedure of referring a dispute to a third neutral party, an adjudicator, who must be appointed within seven days. Once a dispute has been referred to the adjudicator, the adjudicator must act impartially and may take the initiative to ascertain the facts and the law. The adjudicator must fulfil his or her obligation to reach a decision within 28 days of referral and may extend the period of making a decision by up to 14 days with the consent of the referring party or any further extension agreed by the parties. This process aims to determine a dispute on a temporary basis to enable work to proceed unimpeded and with less likelihood of serious injustice being caused (Cottam, 1998). Even if the decision is not accepted by one of the parties, the parties are obliged to implement the adjudicator's decision. The decision is binding unless and until the dispute is finally resolved by legal proceeding, arbitration, settlement agreement or both parties accept the decision as finally determining the dispute.

It is observed that adjudication is similar to arbitration in that it is a judicial process in which the adjudicator determines the parties' respective rights and obligations under the contract on the basis of evidence presented by the parties (Battersby, 2001). The difference is the procedure in adjudication is much simpler as it is intended to be a quick process similar to mediation. Adjudication is not a condition precedent to arbitration or court litigation. It is statutorily enabled which entitles a party to exercise their rights to invoke adjudication, otherwise the parties may opt for other dispute resolution.

In Malaysia, statutory adjudication is mainly enforced via the Construction Industry Payment and Adjudication Act 2012 (CIPAA 2012). Besides providing a speedy dispute resolution mechanism for the construction industry, the other key features of the CIPAA 2012 are to outlaw the practice of pay-when-paid and conditional payment, to facilitate regular and timely payment, and provide security and remedies for the recovery of payment. Under the CIPAA 2012, adjudication is not a condition precedent to arbitration, litigation or other dispute resolution. It is an entitlement which is statutorily provided in the event a party wishes to invoke adjudication. Once adjudication is initiated, the other party is drawn into it. However, the parties are not prevented from resorting to another dispute resolution process, irrespective of whether the proceedings take place concurrently with the adjudication proceedings. Thus, other dispute resolution mechanisms can co-exist, and complement each other.

The Islamic Perspectives of ADR

The Islamic principle which is equivalent to ADR is *sulh* or compromise. The word *sulh* literally means: "to end a dispute," (Ibrahim Anis *et al.*, n.d., p. 520) or "to cut off a dispute" (Muhammad al-Sharbini al-Khatib, 1933). Terminologically it refers to an agreement entered into between two disputing parties which results in the termination of the dispute. The Mejelle defines *sulh* as a contract resolving a dispute by consent.[19]

In general, the above terminological definition of *sulh* seems similar to ADR of common law, in embracing arbitral as well as non-arbitral processes.

Another principle of Islamic law which is relevant to ADR is *tahkim*, which literally refers to arbitration. *Tahkim* means "the submission, by two or more parties to a third party, of a dispute to be adjudicated according to Shariah" (Madkur, 1964). *Tahkim* is also defined as the appointment of a person by two disputing parties to decide their dispute (Mahmud, 1980). The third party is known as *hakam* or *muhakkam* (arbitrator), who is an ordinary man but must possess all the qualifications of a *qadi* (judge) (Abd al-Karim Zaydan, 1995). The *hakam* will determine the dispute according to the Shariah whether or not the dispute has yet to come before the court or is already pending before the court. The award of the *hakam* is binding on both the parties. The whole process from the appointment of the *hakam* to the giving of the award is called *tahkim* (arbitration).

Thus, *sulh* as a process of dispute resolution in Islamic law covers every mode of settlement by the disputing parties in order to end a dispute. This mutual settlement may involve a little assistance from a neutral third party, as in mediation and conciliation, or without any assistance of a third party, as in negotiations, or with the assistance of a neutral third party as in *tahkim* or arbitration. The point which is common to all these processes is that the settlement is made without the court's intervention. Though arbitration is given a distinct term (*tahkim*) in Islamic law, this fact does not push it away from under the wide umbrella of *sulh* because both are based on the mutual agreement of the parties, even though in varying measures.

---

[19] *The Mejelle: Being an English Translation of Majallahel-ahkam-i-adliya and a Complete Code on Islamic Civil Law* (n.d.). English Translation by CR Tyser, *et al.* Lahore: Law Publishing Co., Art. 1531.

Legal Texts on the Permissibility of *Sulh* and *Tahkim*

The legal texts on the permissibility of *sulh* and *tahkim* can be found in the primary sources of the Shariah i.e. the Holy Quran and the *Sunnah*. These two principles are dealt with in the Quran in the following verses:

> "If a woman has reason to fear ill-treatment or desertion on the part of her husband, It shall be no offence for them (the husband and wife) to set things peacefully to rights between themselves (in the best possible manner), for peace is better (for them then a divorce)..." (Quran 4: 128)

> and

> "If you fear a breach between the two (husband and wife), appoint an arbiter from his people and another from hers (to make peace between them). If the arbiters (sincerely) wish them to be reconciled, Allah will bring them together again. Truly Allah is Most Knowing, Most Wise." (Quran 4: 35)

> "And if two parties of believers take up arms against each other, make peace between them...". (Quran 49: 9),

> and from the *Sunnah*, it is reported by Sahl bin Sa'd:

> "Once the people of Quba' fought with each other till they threw stones at each other. When Allah's Apostle (pbuh) was informed about it, he said, 'Let us go to bring about a reconciliation between them.'"[20]

Abu Hurayrah also reported that the Holy Prophet (pbuh) said: "Conciliation between Muslims is permissible."[21] In another *hadith*, it was reported from Shurayh al-Qadi that the Prophet (pbuh) said to disputing parties: "Make peace between you."[22]

Based on *ijma'*, Muslim scholars unanimously hold that compromise is lawful due to its benefit of putting off disputes. Wahbah Al-Zuhayli (1989) states that the legal ruling (*hukm*) for compromise is recommendable (*Sunnah*).

---

[20] Al-Bukhari, Muhammad Ibn Ismaiel (1986). *Sahih Al-Bukhari*. English Translation by Dr. Muhammad Muhsin Khan. 6th Rev. Ed., Vol. 3. Lahore: Kazi Publications, p. 534 at para. 858.

[21] Hasan A (1984). *Sunan Abu Dawud: English Translation with Explanatory Notes*. English Translation by Ahmad Hasan, Vol. 3. Lahore: Sh. Muhammad Ashraf, p. 1020 at para 3587.

[22] Muhammad, b. KH (1980). *Akhbar al-Qudar*, Vol. 2. Beirut: 'Alam al-Kutub, p. 39.

Judges in Islamic law are under imperative duty to ask litigants to compromise. This is based on the *hadith* narrated by 'Abdullah bin Ka'ab to the effect that:

> "Abdullah bin Abu Hadrad al-Aslami owed Ka'ab bin Malik some money. One day the latter met the former and demanded his right and their voices grew very loud. The Prophet (pbuh) passed by them and said, "O, Ka'ab," beckoning with his hand as if intending to say, "Deduct half the debts." So Ka'ab took half what the other owed him and remitted the other half."[23]

Based on the above provisions in the Holy Quran and the *Sunnah*, it becomes clear that disputing parties are strongly urged to compromise to settle their disputes either through mutual agreement or according to the decision of an arbitrator, who is appointed by them mutually.

Statutory Provisions Dealing with *Sulh* and *Tahkim* in Malaysia

Sulh and *tahkim* are recognised in Islamic law. The *Mejelle*, which is perhaps the first and best code of the substantive laws in the Muslim world, and mainly based on the Hanafi law, embodies provisions on *sulh* and *tahkim*. Provisions relating to settlement (*sulh* and *tahkim*) are embodied in the 12th book, which contains 41 articles.[24] The provisions relating to arbitration are embodied in Chapter 3 of the 16th book, which contains 11 articles.[25] However these provisions are general in nature and do not provide useful guide to procedural details.

With regard to modern legislation on Islamic law in Malaysia, *sulh* and *tahkim* are recognised as methods of dispute resolution besides litigation. For example, Section 87 of the Shariah Civil Code Enactment 1991 of Selangor ("the Enactment") explicitly encourages parties to employ *sulh* to settle their disputes. The Enactment also provides that *sulh* can be made at any time (Section 125). The court

---

[23] Al-Bukhari (1986), para. 869.
[24] *The Mejelle* (n.d.) Arts. 1531–1571.
[25] *The Mejelle* (n.d.) Arts. 1841–1851.

may record *sulh* on the request of either party (Section 88). With regard to *tahkim*, Section 48 of the Islamic Family Law Enactment 1984 of Selangor [26] allows *tahkim* in *shiqaq*[27] cases.

Conclusion

The Islamic principles of *sulh* (compromise) and *tahkim* reflect the concept of ADR in the common law as they similarly refer to settlement of disputes out of court. In general, we can conclude that compromise appears to come within the folds of ADR. However unlike common law, which has specific terms, such as negotiation, mediation or conciliation and arbitration, conveying definite meanings, Islamic law has a generic term *sulh* under which various ADR processes may be accommodated. So there is no difference between the two legal systems on this point except in terms of nomenclature. Another distinguishing feature between the two systems relates to the basis of compromise. In the civil system, compromise is based on social and psychological needs whereas in the Islamic system, it is based on divine revelations.

In the Malaysian construction industry, there are many Muslim practitioners *vis-à-vis* the capacity as contractors, developers, consultants and clients. Being Muslim, it is more appropriate to adopt the practices which had been verified by Shariah as best practices. In addition, as the practices of Shariah ADR are not much different from the common law ADR, it is highly commendable that the practices of Shariah be adopted in the Malaysia construction industry. As an initial option, Muslim disputing parties could be given a choice either to adopt the Shariah ADR or the conventional practices.

List of Cases and Legislation

Northern Regional Health Authority v. Derek Crouch Construction Company Limited [1984] 1 QB 644

---

[26]This provision is similar to Section 45 of the Administration of Islamic Family Law Enactment 1985 of Terengganu.

[27]Dispute between a husband and his wife, which is likely to lead to a divorce.

Glencot Development v. Benn Barrett [2001] BLR 207
Balfour Beatty v. London Borough of Lambeth [2002] BLR 288
CIDB Building Works 2000 Edition
United Nations Commission International Trade Law (UNCITRAL)
Model Law on International Commercial Arbitration
New Zealand Arbitration Act
Malaysian Arbitration Act 2005
PAM Arbitration Rules
IEM Rules for Arbitration of the KLRCA
ICC Rules of Arbitration
MIArb Arbitration Rules 2000 Edition
CIDB Standard Form of Contract for Building Works 2000 Edition
PAM Mediation Rules
CIDB Mediation Rules
Rules for Conciliation/Mediation of KLRCA
CIArb Mediation Rules
Malaysian Mediation Center Mediation Rules
ICC ADR Rules
Construction Industry Payment and Adjudication Act (CIPAA) 2012
Shariah Civil Code Enactment 1991 of Selangor
Islamic Family Law Enactment 1984 of Selangor
Convention of the Recognition and Enforcement of Foreign Arbitral
Awards Act 1985 (Act 320)
Administration of Islamic Family Law Enactment 1985 of
Terengganu

References

Abd al-Karim Zaydan (1995). *Nizam al-Qada'*. Beirut: Muassasah al-Risalah.
Abraham C (2006). Alternative Dispute Resolution in Malaysia. Presented at the 9th General Assembly of the Asean Law Association in 2006. http://www.aseanlawassociation.org/9GAdocs/w4_Malaysia.pdf [November 4, 2009]
Al-Bukhari, Muhammad Ibn Ismaiel (1986). *Sahih Al-Bukhari*. English Translation by Dr. Muhammad Muhsin Khan. 6th Rev. Ed., Vol. 3. Lahore: Kazi Publications.
Bagshaw, D (2008). Keynote address in at the 4th Asia-Pacific Mediation Forum 2008 Conference, hosted by the Harun M Hashim Law Centre, at the International Islamic University Malaysia. Kuala Lumpur: June 16–18, 2008.

Battersby, J (2001). Developing trends in construction dispute resolution. In *Arbitration*, Selection of Papers on Arbitration presented at the joint Bar Council/CIArb Talk.Malaysia: The Chartered Institute of Arbitrators (Malaysia Branch).

Brown, H and A Marriot (1999). *ADR Principles and Practice*, 2nd Ed. London: Sweet and Maxwell.

Cheung, SO (2006). Mandatory use of ADR in construction — A fundamental change from voluntary participation, Editorial, *Journal of Professional Issues in Engineering Education and Practice* © ASCE.

Cottam, G (1998). *Adjudication Under the Scheme for Construction Contract*. London: Thomas Telford.

Financial Mediation Bureau. http://www.fmb.org.my [December 23, 2009].

Gaitskell, R (2007). International statutory adjudication: its development and impact. *Construction Management and Economics*, 25, 77.

Government of Malaysia (2007). Standard Form of Contract to be Used Where Bills of Quantities Form Part of the Contract. Public Works Department (PWD) Form 203.

Government of Malaysia (2010). Standard Form of Contract to be Used Where Bills of Quantities Form Part of the Contract. Public Works Department (PWD) Form 203A (Rev. 1/2010).

Groton, JP, RA Rubin and B Quintas (2001). A comparison of dispute review boards and adjudication. *International Construction Law Review*, 18(2), 277.

Hasan, A (1984). *Sunan Abu Dawud: English Translation with Explanatory Notes*. English Translation by Ahmad Hasan, Vol. 3. Lahore: Sh. Muhammad Ashraf.

Hibberd, P and P Newman (1999). *ADR and Adjudication in Construction Disputes*. Oxford: Blackwell Science Ltd.

Holtham, D, V Russell, D Hird, and R Stevenson (1999). *Resolving Construction Disputes*, Oxford: Chandos Publishing.

International Chamber of Commerce. http://www.iccwbo.org/uploadedFiles/Court/Arbitration/other/adr_rules.pdf [Accessed January 7, 2010]

Ibrahim, A, Abdul Halim Muntasir, Atiyyah al-Saralihi, and Muhammad Khalfullah Ahmad (n.d.). *Al-Mu'jam al-Wasit*, Vol. 1, p. 520. Qatar: Idarah Ihya' al-Turath al-Islami.

Khutbul Zaman, Bar Council Alternative Dispute Resolution Committee Chairman in Mediation system to tackle cases fast taking shape, *New Straits Times*, July 3, 2009.

Lim, PG (2004). Mediation, a slow starter in alternative dispute resolution. *Malayan Law Journal* , 1, xv–xix.

Lim, PG and G Xavier (2002). Malaysia. In *Dispute resolution in Asia*, 2nd Ed. M Pryles (ed.), The Hague: Kluwer Law International.Lim, PG (2004). Mediation, a slow starter in alternative dispute resolution. *MLJA*, 1(15).

Mackie, K, D Miles, W Marsh, and T Allen (2000). *The ADR Practice Guide*, 2nd Ed. London: Butterworths.

Madkur, MS (1964). Al-Qada' fi al-Islam. In *Commercial Arbitration in the Middle East: A Study in Shari'ah and Statute Law*, Samir Salleh (ed.), pp. 20–21. London: Graham & Trotman.

Mahmud, b. M b. A (1980). *Tarikh al-Qada' fi al-Islam.* Cairo: Maktabah al-Kulliyyah al-Azhariyyah.

Malaysia Institute of Architects (Pertubuhan Arkitek Malaysia — PAM) (2006). Agreement and Conditions of PAM Contract 2006 (Without Quantities).

Mediation system to tackle cases fast taking shape (July 3, 2009). *New Straits Times*, p. 15.

Kumaraswamy, MM (1998). Consequences of construction conflict: A Hong Kong perspective. *Journal of Management in Engineering*, 14(3), pp. 66–74.

Muhammad al-Sharbini al-Khatib (1933). *Mughni al-Muhtaj*, Vol. 2, p. 177. Cairo, Egypt: Syarikah Maktabah wa Matba'ah Mustafa al-Babi al-Halabi wa Awladuh.

Muhammad, b. KH (1980). *Akhbar al-Qudah*, Vol. 2. Beirut: 'Alam al-Kutub.

Noor Azian, S (2003). Tribunal for consumer claims: The Malaysian experience. Presented at Sixth Annual AIJA Tribunal Conference, June 5–6, 2003. http://www.aija.org.au/Tribs03/Malaysia.pdf [Accessed January 6, 2010].

Patterson, S and G Seabolt (2001). *Essentials of Alternatve Dispute Resolution*, 2nd Ed. Dallas: Pearson Publications Company.

Pẽna-Mora, F, CE Sosa, and DS McCone (2003). *Introduction to Construction Dispute Resolution.* New Jersey: Pearson Education Inc.

Premaraj, B (2007). Arbitration the preferred alternative. Paper delivered at the Asia Pacific In House Congress June 2007.

Rozina Mohd Zafian (2013). Legal analysis of statutory adjudication in the construction industry: Special reference to UK and Malaysian acts. Unpublished PhD Thesis. International Islamic University Malaysia.

Stevenson, R and P Chapman (1999). *Construction Adjudication.* Bristol: Jordan Publishing Limited.

Sundra Rajoo and WSW Davidson (2007). *The Arbitration Act 2005 — UNICITRAL Model Law as applied in Malaysia.* Petaling Jaya: Sweet & Maxwell Asia.

Sykes, JK (1996). Claims and disputes in construction: Suggestions for their timely resolution. *Construction Law Journal* 12(1), 3–13.

The Mejelle: Being an English Translation of Majallahel-ahkam-i-adliya and a Complete Code on Islamic Civil Law (n.d.). English Translation by CR Tyser, *et al.* Lahore: Law Publishing Co.

Treacy, TB (1995). Use of alternative dispute resolution in the construction industry, *Journal of Management in Engineering*, 11(1), 58–63.

Uff, J (2005). *Construction Law*, 9th Ed. London: Sweet and Maxwell.

Wahbah al-Zuhayli (1989). *al-Fiqh al-Islami wa Adillatuh*, 3rd Ed., Vol. 5. Damascus: Dar al-Fikr.

Xavier, G (2002). Comparative study of arbitrations in Malaysia and selected jurisdictions in the European Union. *The Malayan Law Journal*, lxxxix–cxxi

Part 3

Shariah-Compliant Project Finance and Risk Management

Chapter 10

The 3 Rs in Islamic Project Finance: Its Relevance Under *Maqasid al-Shariah*

Etsuaki YOSHIDA

Introduction

Islamic finance has enjoyed stable growth till today in its commercial practice. It shows geographical spread, not just within the Middle East and Southeast Asia, but also to emerging economies such as South Asia, Commonwealth of Independent States (CIS) region and Africa, and even to Muslim-minority countries such as the United Kingdom, Singapore, Luxembourg, France, Japan, Hong Kong and South Africa.

The current market size is estimated around USD 2 trillion or more, with a stable growth rate around 15% a year. While the origin of the modern Islamic finance is considered to be the commercial success of Dubai Islamic Bank (established in 1975), the banking asset occupies 81% of the world's total Shariah-compliant financial assets, according to GIFF (2012).

Prohibition of *riba*, or generally perceived as interest rate, is one of the most outstanding features in Islamic finance, especially so for non-Muslims. Needless to say, the prohibition of *riba* is stated in several verses in the Holy Quran, e.g.:

> "... Whereas Allah has permitted trading and forbidden usury..."
> (Quran 2: 275).

> "As for those who take usury, they shall not be able to stand upright but
> shall rise up like one whom Satan has demented by his touch..." (Quran
> 2: 275).

> "O believers! Remain conscious of Allah and give up what is still due to
> you from usury (from those who are still indebted to you) if you truly
> believe. But, if you do not do so (by disobeying the prohibition of usury),
> then be warned of the war that shall be declared against you by Allah and
> His Messenger (which has terrible consequences). And if you repent now,
> you may retain your principal (without interest). (Consequently,) deal not
> unjustly, and you shall not be dealt with unjustly." (Quran 2: 278–279).

Entrepreneurs, bankers, lawyers and Shariah scholars in the
Islamic financial industry have successfully developed various
types of Islamic financial products, which are commercially acceptable to both financial service providers and customers, and in
line with Shariah principles. The biggest attention has been paid
to how to structure *riba*-free financial products which have the
same economic function with each conventional equivalent, as
seen in Islamic deposit, Islamic auto-loan, Islamic home financing,
Islamic trade finance, Islamic interbank money market fundraising,
sukuk, "profit-rate" swap and so on. Several conceptual models
were invented to achieve Shariah-compliant debt products, such as
*Murabaha, Tawarruq, Ijara, Ijara Muntahia Bitthamleek, Istisna', Salam,
Musharaka Mutanaqisa* and so forth.

Project financing is not an exception. The USD1.8 billion
Hub River project in Pakistan in 1994 is understood as the first
Shariah-compliant project financing deal (Dar, 2010), followed by
the Kuwaiti Equate petrochemical project in 1996. Many transactions were made since then, and Islamic Project Finance, or
hereafter IPF, is not a rare and novel technique for today's
financiers.

This chapter discusses the nature of IPF and "profit rate" in it,
which is the Islamic financial term for interest rate, mainly from the
viewpoint of risk-sharing.

Desirable Direction of Development of Islamic Finance

A Theoretical Approach Under Shariah

Before moving on to the discussion on project finance, let us discuss a desirable direction of Islamic finance development under Shariah, to have a better understanding of "profit rate" in IPF.

Principles of Shariah basically prefer "equity-based" transactions to "debt-based" ones. The major supporters of this idea were Islamic economists such as Umer Chapra, Nejatullah Siddiqi and Osman Ahmed, who were called the "Jeddah School" by Zubair Hasan (2005). Nagaoka (2012) called this widespread preference of equity-based finance among Islamic economists as "*mudaraba* consensus."

However, how can we assess the actual development of Islamic financial products from a viewpoint of this *mudaraba* consensus? On the product front, practitioners in the financial industry have developed various types of Islamic products, and now it is not too much to say that the product suite of Islamic finance is roughly similar to that of the conventional equivalent. It has continued to evolve, mainly in the direction of realizing the same function with conventional instruments, by skillful arrangement of financial and legal techniques.

Figure 10.1 shows the brief history of product development of Islamic finance in chronological order.

However, from the viewpoint of *mudaraba* consensus, the direction of the development of IPFs is not toward pursuing the objective of the religion, or *Maqasid al-Shariah*. Actually, GIFF (2012) indicates that 93.4% of Islamic financial assets is debt-based. Such being the case, practitioners (and some other economists) tend to consider their idea on product development as mentioned above natural, implying that the equity-based financial system, which Islamic economists prefer, is not realistic. Gainor (2000) describes this recognition in a very concise manner. "Much of the research and development that has worked its way into existing products in the marketplace has been generated from adapting conventional products. It may follow that if a product was successful in the

Year	Main area of product development in its first deal
1950–1963	Prototypes of financial institutions [South Asia, Egypt, Malaysia]
1975–1979	Genuine practice of banking activity [Middle East, North Africa]
1979	*Takaful* [Sudan]
1986–1993	Equity funds [US, Singapore, South Africa, etc.]
1990	*Sukuk* [Malaysia]
1994	Project finance [Pakistan]
2005	Securitized (residential mortgage-backed) product [Malaysia]
2006	Exchange-traded fund (ETF) [Turkey]
2006	Derivatives (profit-rate swap) [UAE, Malaysia]

Figure 10.1.   Chronological Development of Major Islamic Financial Products.
Source: Collated by the Author Using Various Sources.

conventional marketplace, then if successfully engineered as to not be inconsistent with Islamic Shariah, it should be successful in the Islamic marketplace." In short, Shariah theoretically prefers development of equity-based financial transactions, while practitioners remain debt-oriented as in the conventional financial system.

Critiques by Islamic Economists

Although the Islamic finance industry has shown remarkable growth as sketched in the previous section, the current situation is not necessarily welcomed by Islamic economists (in this chapter, the definition of "Islamic Economists" and "Islamic Economics" are simply scholars that deal with economic issues in consideration of religious values of Islam, and the academic approach by them, respectively). There is wide agreement among them that the majority of Islamic financial markets, or assets, should not be occupied by debt-based transactions, such as *murabaha* and *ijara*, as it actually is today. This tendency is concisely represented by the phrase, "*Murabaha* Syndrome," a terminology by Tariq (2004).

On top of that, Chapra (2007) argues that the share of equity-based transactions should increase in the current financial system, while that of the debt-based ones should decrease substantially. El-Gamal (2003) described the current situation as "Islamic finance quickly turned to mimicking the interest-based conventional finance." In addition, Hamoudi (2007) called the current situation "Jurisprudential Schizophrenia" and De Lorenzo (2007) bantered it as "Shariah-conversion technology." Ahmed (2011) observed the situation in a more objective manner, saying "Contemporary practice of Islamic finance has been criticized for not fulfilling the *maqasid*."

To sum up, although Islamic finance in its religious theory prefers equity, the current practice of the industry remains debt-oriented as the conventional financial system is. Efforts have so far been made to get closer to the religious objectives. In Malaysia, Bank Negara Malaysia took the leadership in establishing Investment Account Platform (IAP),[1] which enables account holders of member Islamic banks to invest in a specific project, the details of which are on the IAP website. IAP provides detailed information on the project, and an individual may invest in the equity. It was launched in 2015, and expected to expand.

However, in reality, although "dividend" or investment return is preferred, mere "profit-rate" as the interest rate equivalent (and a nomenclature) prevails in the actual markets of Islamic finance.

Considering 3Rs: Riba, Recourse, Risk-Sharing

Project Finance and Interest

While often used, "project finance" looks misunderstood especially among people outside the banking industry. It does not simply mean "finance (or loans) to a (big) project." Rather, the word

---

[1] For the modus operandi of IAP, see its website at https://iaplatform.com/showI apInfo [Accessed November 3, 2018].

"project finance" in the technical term within the banking industry should be understood as "non-recourse (or limited-recourse) finance."[2] It can be interpreted as the antonym of "full recourse finance."[3]

The concept of full recourse finance is simple (see Figure 10.2). It is something that many people regard as a loan. When a loan is extended to a borrower from a lender (the amount is represented as X in Figure 10.2), the borrower must repay the obligation of the principal and the interest on a specified repayment date. In other words, the credit risk of the borrower is contained in this transaction. This is one of the reasons why *riba* is prohibited in Islam: the borrower must be engaged in a business, taking all the risks accompanying the business, while the lender is completely sure to gain profit as interest, as stipulated in the loan agreement, without taking any risks of the business, and hence, the situation of the two parties is unfair. Under Shariah, this is against "*'adl*," meaning fairness or justice. The obligation remains even if the borrower fails to make enough profit out of the business, which was expected to be the source of repayment. If the borrower fails to repay the debt, described as "falling into default," the lender will have to seek recourse from the borrower for recovery of the claim.

"Non-recourse finance," on the contrary, means that the borrower itself is NOT the target of recourse when in default. Basically, the source of repayment is the pool of cash flow (shown as "Escrow Account" in Figure 10.2) generated by the project constructed with the finance extended, not the borrower itself. To state it differently, the risk lies in the project, and in this sense, the lenders take the risk of the project, as well as the borrower: if the project fails to generate enough profit, the lenders will not get repaid in full amount on the specified repayment date. In this regard, this scheme can be interpreted as profit/loss sharing contract between a lender and a borrower. Meanwhile, the borrower as the project sponsor will

---

[2]Limited-recourse refers to a situation where the project sponsor takes some recourse.

[3]For basic understanding of project finance, see Finnerty (2007) as one of the standard textbooks in this category.

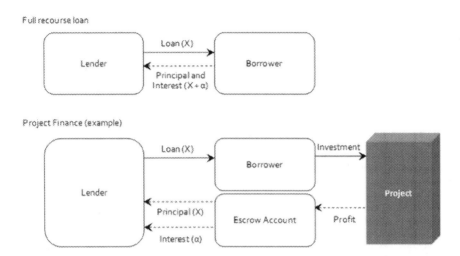

Figure 10.2.   Full-Recourse and Non-Recourse Loans.
Source: Author.

concentrate on the project business, as they have no obligation of repayment, under the financing agreement, if the project does not succeed.

In this regard, McMillen (2000) defines the term as follows: "In general terms, a 'project financing' is the financing of an economic unit in which the lenders look initially to the cash flows from operation of that economic unit for repayment of the project loan and to those cash flows and the other assets comprising the economic unit as collateral for the loan."

Classifying Financial Products — Comparative Position of IPF

Islamic Project Finance contains elements of both "non-recourse" and Shariah compliance by definition. Shariah compliance is achieved typically using contract modes of *Istisna'* and *Ijara maw-sufah fi al-dimmah* (forward lease agreement) for infrastructure and petro-related projects. Non-recourse nature is met using the same structure with conventional project finance by establishing a project company and involving security package including collaterals. In this context, it is noteworthy that (Islamic) project finance contains elements of "risk-sharing" regarding the project. The lenders in a project finance deal usually bear risks of the project; if the operation of the project does not go appropriately, the lenders may not get repaid in full amount on a specified due date. Practically, the lenders tend to secure their claims by shifting the risk to other parties: using insurance services and involving guarantors such as governments.

To understand the nature of IPF in a more comprehensive manner, Figure 10.3 shows comparison with other financial products according to existence of factors, which are essential to Islamic finance.

In Figure 10.3, equity transaction is included in the coverage of comparison, which is more desirable than debt transaction in terms of Islamic principles as discussed earlier. Also, non-recourse and risk-sharing is roughly two sides of the same coin, so they are in the same column.

It is obvious that (i) conventional full recourse finance has nothing good for Shariah: that is why they are *haram* (prohibited) and should be structured in some ways in order to be Shariah-compliant. (ii) Islamic full recourse finance is *halal* (permissible) in the sense that the *riba* factor is eliminated by applying Shariah-compliant

		Non-*riba*	Non-recourse/ risk-sharing	Equity
1)	Conventional/ Full recourse	No	No	No
2)	Islamic/ Full recourse	Yes	No	No
3)	Conventional/ Non-recourse	No	Yes	No
4)	Islamic/ Non-recourse	Yes	Yes	No
5)	Equity	Yes	Yes	Yes

Figure 10.3.   Comparison of Financial Products by Factors.
Source: Author.

financial structure. However, it is also a fact that this type of financing is often criticized as "Shariah conversion technology" and "mimicking of conventional equivalent." (iii) Conventional non-recourse finance would be interesting for academic discussions of Shariah studies: under current practices of dealing with *riba*, it is not Shariah-compliant, but if we consider the form and content of finance, it is much more desirable because it has risk-sharing nature and is frequently used for construction of tangible assets, which leads to the concept of *"'ain."*[4] (iv) Islamic non-recourse finance contains both factors of non-*riba* and non-recourse as discussed earlier. Although it is behind equity deals in the order of preference under Shariah, we can see clear differences from other types of debt-based finance.

In this sense, the relevance of IPF should be understood and stressed more in academic studies of Islamic finance. When we see past studies of IPF, there is a tendency that authors of these works are limited to those with professional legal backgrounds[5] with little contributions from Shariah studies and Islamic economics.

Hypothetical Proposals and Conclusion

As discussed, IPF contains elements of risk-sharing, which is in line with the principles of Shariah, although it may look like mere "mimicking" of the conventional equivalent to the eyes of some Islamic economists. In this sense, IPF can be recognized as a "better" Islamic financial product in the degree of *Maqasid al-Shariah*.

The author would like to propose two hypotheses for more lively discussions of IPF in the context of Islamic economics and Shariah studies, partly in an effort to live up to the hopes of Islamic economists.

---

[4]The Arabic term *'ain* is tangible asset, while *dayn* is its antonym.
[5]For example, McMillen (2000) [professional lawyer], Alexander (2011) [law school student] and Sadikot (2012) [law school student]. Babai (1999) and Lee and Son (2013) look from the discipline of regional studies and economic research, and they do not refer enough on Shariah aspects.

Hypothetical Proposal 1:

> *All banking institutions offering Islamic financial services should adopt "Islamic non-recourse scheme" in their debt products as much as possible, to get closer to the religion's ideal form of finance.*

As discussed in the previous section, normal Islamic financing, or Islamic full-recourse finance, is less desirable than IPF. Usually, project finance requires a lot more costs of structuring a deal, including legal service fees, than full recourse equivalents, but there will be some other ways, e.g. using legal techniques or banks' business attitudes for even smaller deals such as home financing. This will make a big difference from conventional banks.

Hypothetical Proposal 2:

> *Interest rate in conventional project finance should be reconsidered as Shariah-compliant as it contains the nature of risk-sharing.*

Interest is prohibited in Islam, as mentioned in the first section of this paper citing some verses of the Holy Quran. However, if we consider the reasons and the details of what is prohibited as *riba*, it is academically significant to deliberate the possibility of not regarding the "profit margin" of conventional project finance as *riba*, since it is a fixed amount of the profit from the project of which the lenders take risk.[6] There does not seem to be much discussion done on this point among Shariah scholars and Islamic economists. If this ambitious hypothesis is deemed legitimate, efficiencies of project finance transactions will increase, and hence, there will be a possibility of enhancing project finance in many other areas of global finance. It will lead to economic development, which is the objective of Islamic finance.

To conclude, a lot of debates have been done regarding *riba*, in a situation where majority of interest rates is latently understood in the form of full recourse finance. As project finance develops and gains more popularity, there will be higher possibility of considering those hypotheses mentioned earlier.

---

[6]Technically speaking, the interest here may be synthetically interpreted as *"ex ante* fixed profit from *mudaraba sukuk."*

Project finance is for construction of tangible (and productive) assets, which is desirable under Islamic economics. It is also desirable because the project will contribute to economic development of a country, which may lead to poverty alleviation that contributes to *Maqasid al-Shariah*. More discussion on this point among Shariah scholars and Islamic economists is expected.

References

Ahmed, H (2011). Maqasid al-*Shariah* and Islamic financial products: A framework for assessment. ISRA *International Journal of Islamic Finance* 3(1), 149–160.

Alexander, AJ (2011). Shifting title and risk: Islamic project finance with Western partners. *Michigan Journal of International Law*, 32(3), 571–612.

Babai, D (1999). Islamic project finance: Problems and promises. In *Proc. Second Harvard University Forum on Islamic Finance: Islamic Finance into the 21st Century*.

Chapra, MU (2007). The case against interest: Is it compelling? *Thunderbird International Business Review*, 49(2), 161–186.

Dar, H (ed.) (2010). *Global Islamic Finance Report 2010*, London: BMB Islamic UK.

De Lorenzo, YT (2007). *The Total Returns Swap and the "Shariah Conversion Technology" Stratagem*. New York: Dinar Standard.

El-Gamal, MA (2003). "Interest" and the paradox of contemporary Islamic law and finance. *Fordham International Law Journal*, 27(1), 108–149.

Finnerty, JD (2007). *Project Financing: Asset-based Financial Engineering*, Second Ed. Hoboken: John Wiley & Sons.

Gainor, T (2000). A practical approach to product development. Paper prepared for the Fourth Harvard University Forum on Islamic Finance.

Global Islamic Finance Forum (GIFF) (2012). *Islamic Finance Opportunities: Country and Business Guide*. Kuala Lumpur: KFH Research Ltd.

Hamoudi, HA (2007). Jurisprudential schizophrenia: On form and function in Islamic finance. *Chicago Journal of International Law*, 7(2), 605–622.

Investment Account Platform. https://iaplatform.com/showIapInfo [Accessed November 3, 2018].

Lee, KH and SH Son (2013). Trends and implications of Islamic project finance: A study on the GCC region world economy update, Vol. 3, No. 24, Korea Institute for International Economic Policy.

McMillen, MJT (2000). Islamic Shariah-compliant project finance: Collateral security and financing structure case studies. *Fordham International Law Journal*, 24(4), Article 6.

Nagaoka, S (2012). Critical overview of the history of Islamic economics: Formation, transformation, and new horizons. *Asian and African Area Studies*, 11(2), 114–336.

Sadikot, R (2012). Islamic project finance: Shari'a compliant financing of large scale infrastructure projects. *Al Nakhlah Online Journal on Southwest Asia and Islamic Civilization*, Spring, 1–9.

Tariq, MY (2004). The Murabaha syndrome in Islamic finance: Laws, institutions and politics. In *The Politics of Islamic Finance*, CM Henry and R Wilson (eds.), pp. 63–80. Edinburgh: Edinburgh University Press.

Zubair Hasan (2005). Islamic banking at the crossroads: Theory versus practice. In *Islamic Perspectives on Wealth Creation*, M Iqbal and R Wilson (eds.), pp. 11–25. Edinburgh: Edinburgh University Press.

Chapter 11

Islamic Home Financing Through *Musharakah Mutanaqisah*: A Crowdfunding Model

AHAMED KAMEEL Mydin Meera

Introduction

Having a shelter called home is a basic necessity of life. Everyone needs a shelter to take protection from sun, rain, harsh weather and to take rest, sleep and do other private things. It is a place to dwell in tranquillity with family and raise children. Therefore, owning a good home is everyone's aspiration. Most people of today fulfil this need by purchasing a home from developers and financing the purchase by taking a mortgage from the bank. Interest-based conventional home mortgages are forbidden in Islam. Therefore, Islamic banks, instead of giving loans, would purchase the home and sell it to the buyer with a profit mark-up that is generally repayable within a long duration of time, i.e. 25 to 35 years.[1] Indeed, financing a home through conventional or Islamic modes normally takes a good chunk of one's monthly income, as much

---

[1] As is the case in Malaysia.

as 25% to 40%. Very few people nowadays build their own homes without taking any financing from outside. The price of homes are skyrocketing in most countries predominantly due to the interest-based fiat monetary system, thereby making homes increasingly unaffordable to many people.

Islamic financial institutions use a number of Shariah-compliant contracts to finance home ownership, particularly the *al-Bai Bithaman Ajil* (BBA) and the *Musharakah Mutanaqisah* (MM) contracts. Conservative Shariah scholars like those in the Middle East disapprove of the BBA contract, citing that it is similar to conventional loans. The BBA contract is a long-duration *murabahah* contract, i.e. a sale with a mark-up that is repayable monthly for, say, 35 years. This contract has proven to be a "disaster" in the Malaysian case, the reason for which is discussed later in this chapter. Theory of finance also predicts the BBA contract to converge to conventional interest-based loans due to the *law-of-one-price* (Meera and Larbani, 2004). This convergence has caused Islamic finance to evolve greatly; from being a "profit-and-loss" banking to a floating rate financing mode which mimics conventional finance. Indeed, the very thing that Islamic finance seeks to avoid — the interest rate — becomes the benchmark for Islamic profit rates.

The objective of this chapter is to illustrate an alternative Islamic home financing scheme using equity methods that make financing the purchase of a home comfortable within a shorter duration, i.e. within an average of 10 years only. Since economic and environmental sustainability is a hot topic nowadays due to drastic climate changes observed and destruction to ecosystems, this chapter also suggests the building of homes having sustainable features. These can indeed be regarded as Shariah-compliant construction.

The next section discusses the monetary sector and the reasons for the drastic increase in home prices that make homes unaffordable. This could help the reader understand why our proposed model can indeed go counter to interest-based fiat money effects and thereby check home price increases.

The Monetary System — The Cause for Drastic Increase in Home Prices

Monetary economist Milton Friedman said that inflation is predominantly a monetary phenomenon. Inflation can also happen when demand continuously exceeds supply, or for some reason supply continuously shrinks. However now many economists echo Milton Friedman's statement that inflation is a direct product of the structure of the monetary system itself.

For example, Lietaer (2001), Greco (2001), Ahamed Kameel (2002, 2004) and many other scholars have identified the nature of money and the monetary structure as the root cause of inflation and many other socio-economic consequences. The main features identified as contributing to inflation are (1) fiat money, (2) Fractional Reserve Banking (FRB) system and (3) compound interest. These features are global and found in almost every modern economy. The issue with these three characteristics is that they cause the money supply in the economy to grow continuously at an exponential rate, exceeding that of the real economy. Firstly, because fiat money is not backed by anything real like gold. It is created by simply printing currency money or by the stroke of a pen by the banking system when it creates loans. The FRB, on the other hand, multiplies bank deposits through the fractional reserve ratio. If the reserve ratio is say 10% and someone deposits $1,000,000 into the bank, the banking system can create loans to the amount of $9,000,000 so that the total deposits become $10,000,000. Now the original deposit of $1,000,000 is 10% of total deposits as required by the reserve requirement. The third component, i.e. compound interest now increases the money supply continuously through the charging of interest on the newly created loans. This growth is exponential because the mathematical formula for compound interest is exponential.

The reason why it is postulated that compound interest is significantly at the root of the global chaos of today is that under the present fiat monetary system, most money is created through mere accounting entries by commercial banks, created in the form of loans, that carry compound interest with it. Due to this reason,

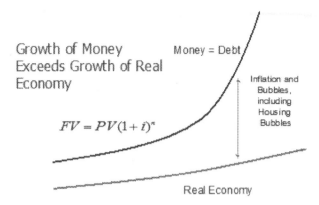

Figure 11.1. The Mismatch Between Monetary Growth and Real Economy Growth.

both money and debt grow exponentially within the economy. But nonetheless the real productive economy, i.e. the economy that produces goods and services, does not and cannot match the exponential growth of money and debt. This is where root of the whole global problem lies (see Figure 11.1).

The structure of the monetary system (shown in Figure 11.1) is hence responsible for the ever-increasing inflation and the creation of the poor and destitute in the economy. It is also responsible for the ever-increasing price of homes, thereby making home-affordability beyond the reach of many.

As a consequence, money and debt overshoot the real economy, the difference of which shows up in the form of inflation and bubbles, particularly stock market and housing bubbles. In the process, individuals, businesses and governments become increasingly indebted (see Table 11.1). When the average debt level reaches a point that is unbearable for the average entity in the economy, it then "bursts," causing an avalanche of foreclosures of properties and the "destruction" of money in the process, which in turn brings about economic recessions, unemployment and so forth.

Hence it is clear that housing bubbles that cause homes to become unaffordable to many is significantly caused by the monetary system.

Table 11.1.   Total Global Debt in Selected Countries and
Economic Groups.

Global Debt, 2001–2016 (trillions of USD)			
	2001	2007	2016
TOTAL	62	116	164
Advanced Economies	55	100	119
United States of America	20	34	48
Japan	13	16	18
Emerging Markets	6.4	16	44
China	2	5	26
Low-income Developing Countries	0.3	0.5	1.3

Source: IMF, Fiscal Monitor April 2018.

In the conventional system, home financing is, of course, usually interest-based and forbidden in Islam. The current BBA home financing, however, does not alter much of the above equation. Instead of charging the customer interest, financiers charge a profit derived through a buy-and-sell contract which is permitted in Islam,[2] but regretfully, the profit rate is dependent on the market interest rate due to arbitrage activities.[3] Therefore, while the BBA is practiced as Shariah-compliant in some countries, it is, nonetheless, converging to the conventional mode where the computational formulas are similar to the conventional and where the profit rate tracks the market interest rate.

BBA Home Financing and Its Operational Issues

BBA is a facility provided by the financier to assist the customer to pay the cost of financing, e.g. a home, over the tenure of financing, e.g. 30 years, at a fixed rate determined by the financier.

---

[2]This is based on the frequently quoted verse from the Holy Quran 2: 275 which states that Allah has permitted trade and forbidden *riba*.

[3]Which Shaikh Nizam Yaquby and Muhammad Taqi Usmani, in a 1998 *fatwa*, approve of so long such interest-based benchmarks do not render the contract on a variable rate.

The financier initially buys the house from the customer (cost of financing amount) and sells it back to the customer, plus its profit margin.[4]

As the seller of the home, Shariah requires the bank to hold ownership of the property and to hold all liabilities arising, including defects. But currently, BBA documentations show that the bank merely acts as a financier rather than a seller and excludes itself of all liabilities. This, of course, ignores the Shariah principle of *"al-Ghorm bil Ghonm"* (no reward without risk), *"Ikhtiar"* (value-addition or effort) and *"al-Kharaj bil Daman"* (any benefit must be accompanied with liability), thereby rendering the BBA profit to be implicated with *riba*.

Saiful Azhar (2005) also opined that there is no risk-taking in the current BBA financing, and, hence, does not merit the Quranic concept of *al-bai*. *Daman* (liability) should also exist in a trading transaction whereby the supplier provides guarantees on the goods sold. However in the current BBA home financing, the customer is forced to face the financial burden of paying for the house even before it is completed, as he has engaged in a "debt contract" with the bank at the outset. By ignoring the concept of *iwad*, the BBA contract is not seen as conforming to the *Maqasid al-Shariah* that removes hardship (*raf' al-haraj*) and preventing harm (*daf' al-darar*) in the economic sphere, thereby leaving the welfare of people unprotected — a possible crime when the transaction is done under an Islamic label (Bilal Bin Farid *et al.*, 2005).

---

[4]Indeed, therefore the BBA generally also involves a *bay al-Inah* contract that is disapproved by all schools except the Shafi'e school that permitted it with abhorrence (see Mustafa Omar Mohammed (2005), p. 17). Furthermore, it is important to note that most if not all Islamic banks operate under the fractional reserve system. This means that Islamic banks, just like their conventional counterparts, also create fiat money out of nothing but disburse this newly created money using Islamic financing modes. Ahamed Kameel and Moussa (2004) and others have questioned the Shariah compliance of fiat money creation itself, an issue on which most Shariah scholars are still silent about.

Example of al-Bai Bithaman Ajil (BBA) Financing

Consider the following example.[5] Assume that a customer wishes to buy a house priced at RM400,000. The customer puts a down-payment of 10%, i.e. RM40,000 and finances the remaining 90%, i.e. RM360,000 using the BBA method. Also assume that the Annual Percentage Rate (APR) charged by the bank is 10% per annum and the duration of financing is for 30 years. The Islamic bank would first buy the house for RM360,000 and then sell the house to the customer at a profit, with deferred payments over the 30-year period.[6]

The monthly payment for the above financing is RM3,159.26,[7] payable for 360 months which adds up to RM1,137,332.76 in total. The difference between this figure and the original financing of RM360,000 which equals RM777,332.76 is the total profit for the Islamic bank from this transaction. The profit of RM777,332.76 is capitalized upfront in the BBA mode, unlike under the conventional mortgage, where the interest due is not recognized until the lapse of time. One important difference of the BBA compared with the MM and the conventional mortgage is that of the balance of financing remaining before the expiry of the duration of financing. For example, the BBA balance after 15 years (i.e. after 180 payments) is the total of the remaining 180 payments, i.e. RM568,666.38 whereas under conventional mortgage, the balance is this amount less the total interest to be paid for the loan over the remaining 15-year period, i.e. only RM293,992.43. However, nowadays the Islamic bank is required to give some rebate for early redemption, the formula for which is given by the central bank.[8] Note that even after

---

[5] A condition that existed in a locality in Malaysia in 1996.

[6] It is important to note that Islamic banks and conventional banks that operate under the fractional reserve system would, indeed, create this RM360,000 out of nothing. This mechanism actually adds new money into the economic system without reducing the total deposits of the depositors.

[7] Computed using the standard formula for present value of annuities, i.e.

$$PV = \frac{Pmt}{i}\left[1 - \frac{1}{(1+i)^n}\right] \quad \text{which gives } Pmt = \frac{i(1+i)^n PV}{(1+i)^n - 1}.$$

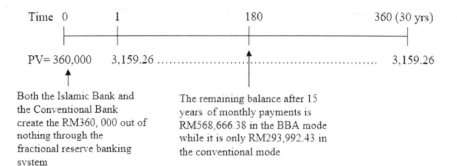

Figure 11.2. Islamic Banking and Conventional Banking Under Fractional Reserve System.

15 years of repayment, the balance under the BBA mode can even exceed the original financing of RM360,000[9] (see Figure 11.2). Nevertheless, considering all the socio-economic effects of fiat money, we regretfully assert that Islamic banking within the fractional reserve system can, indeed, be very damaging to the economy.[10]

While both the Islamic bank and the conventional bank create the original principal amounts through fractional reserve banking system (i.e. loans given out do not really reduce the deposits of the depositors), a customer owes more money in the Islamic mode

---

[8]In sale contracts, the Shariah actually prohibits the rebate to be stated as part of the contract.

[9]This happens neither under the conventional mortgage nor the MM as will be illustrated later, where the balance of financing, at any point in time, never exceeds the principal amount. Therefore, under fractional reserve banking system where both the conventional and Islamic banks create money out of nothing, the Islamic mode is, indeed, very attractive to the bankers. In fact, Citicorp, a conventional bank, is reported as the largest Islamic banking services provider in the world in terms of transactions (*The Asian Wall Street Journal*, May 3, 2005, p. 1).

[10]See Ahamed Kameel (2002, 2004) for a discussion on the socio-economic implications of fiat money creation and for a demonstration why both Islamic and conventional banking systems would ultimately converge, i.e. become equal, due to arbitraging between both the banking systems. This implies that the global Islamic mode of financing would soon converge into the floating rate market. Ahamed Kameel and Moussa (2005) argued that the seigniorage of fiat money is, indeed, a profound *riba*, which the Shariah scholars must address.

than the conventional mode at any time thereafter until the loan is settled. This fact alone is very attractive for even the conventional bankers to provide Islamic mode financing. But, nonetheless, considering the serious negative socio-economic implications of fiat money-based fractional reserve banking, we regretfully conclude that Islamic banking under fractional reserve system is likely to accelerate the said effects — default rates, transfer of wealth and sovereignty, etc. (Ahamed Kameel, 2004).

The *Musharakah Mutanaqisah* (MM) Contract

The *Musharakah Mutanaqisah* (MM) contract, on the other hand, is based on a diminishing partnership concept. Here, there are two portions to the contract. First, the customer enters into a partnership (*musharakah*) under the concept of "*Shirkat-al-Milik*" (joint ownership) agreement with the bank. Customer pays, for example, 10% as the initial share to co-own the house whilst the bank provides for the balance 90%. The customer will then gradually redeem the financier's 90% share at an agreed portion periodically until the house is fully owned by the customer. Second, the bank leases its share (90%) in the house ownership to the customer under the concept of *ijarah*, i.e. by charging rent; and the customer agrees to pay the rental to the bank for using its share of the property. The periodic rental amounts will be jointly shared between the customer and the bank according to the percentage shareholding at the particular times which keep changing as the customer redeems the financier's share. The customer's share ratio would increase after each rental payment due to the periodic redemption until eventually the home is fully owned by the customer.

Boualem and Tariqullah (1995) and Muhammad Taqi (2002) basically agreed on the implementation process of MM. It is best to implement MM for home, vehicle or machinery financing whereby such assets can be leased out according to agreed rental. Joint ownership of assets is accepted by all schools of Islamic jurisprudence since the financier sells its shares to the customer (Muhammad Taqi, 2002). Hence, creating joint ownership in the form of *Shirkah al-Milk* is allowed in the Shariah (Muhammad Imran Ashraf, 2002, p. 116).

Example of a Musharakah Mutanaqisah (MM) Partnership Financing

Consider the same example as the one used for the BBA concept where a customer wishes to buy a house priced at RM400,000. Let us assume again that the customer pays 10% of the price, i.e. RM40,000, and the financier puts the remaining 90%, i.e. RM360,000 and that the average rental for similar homes in the locality is agreed upon between the two parties to be RM2,000 per month. And the customer wishes to add another RM158.38 monthly[11] in order to redeem the financier's share in 30 years. This gives the total monthly payment as RM2,158.38. Table 11.2 below provides the schedule for the MM contract.

Notice that while the amount to be paid monthly was RM3,159.26 under the BBA concept, the monthly amount needed under MM is only RM2,158.38. Therefore, the customer saves RM1,000.88 monthly but acquires the home in 30 years. Indeed, if the customer pays RM3,159.26 for the MM mode as in the BBA, then the customer can own the home in 14 years 2 months,[12] i.e. saving about 15 years and 10 months of monthly payments, i.e. more than half the duration under BBA. Table 11.3 provides a comparison for financing the home using the conventional, BBA and MM methods.

From Table 11.3, it is obvious that so long as the APR are the same, the total interest in the conventional equals the total profit in the BBA. But when customer wants to settle the financing earlier, say after 15 years, the loan balance under the BBA is always higher than under the conventional loan. The balance under the conventional is much lower because here the balance is the present value of the remaining 180 payments whereas under the BBA it is simply the monthly payment times 180 (i.e. under BBA the total profit for the 30 years is capitalized upfront). Nonetheless, the bank nowadays gives a rebate for early settlement, as required by the

---

[11] Equation (3) in the Mathematical Derivation for *Musharakah Mutanaqisah* given in the Appendix is used to obtain this amount of RM158.38.

[12] Obtained using Equation (2) in the Appendix.

Table 11.2. Payments Schedule for *Musharakah Mutanaqisah* Partnership.

Month	Monthly rent (RM) A	Monthly redemption (RM) B	Total payment (RM) C = A + B	Customer's ratio D	Rental division Customer E	Rental division Financier F	Customer's equity (RM) G	Financier's equity (RM) H	Financier's cashflow (RM)
0							40,000.00	3,60,000.00	(3,60,000.00)
1	2,000	158.38	2,158.38	0.10000	200.00	1,800.00	40,358.38	3,59,641.62	2,158.38
2	2,000	158.38	2,158.38	0.10090	201.79	1,798.21	40,718.55	3,59,281.45	2,158.38
3	2,000	158.38	2,158.38	0.10180	203.59	1,796.41	41,080.52	3,58,919.48	2,158.38
4	2,000	158.38	2,158.38	0.10270	205.40	1,794.60	41,444.31	3,58,555.69	2,158.38
5	2,000	158.38	2,158.38	0.10361	207.22	1,792.78	41,809.91	3,58,190.09	2,158.38
6	2,000	158.38	2,158.38	0.10452	209.05	1,790.95	42,177.34	3,57,822.66	2,158.38
—	—	—	—	—	—	—	—	—	—
—	—	—	—	—	—	—	—	—	—
—	—	—	—	—	—	—	—	—	—
—	—	—	—	—	—	—	—	—	—
360	2,000	158.38	2,158.38	0.99463	1,989.25	10.75	3,99,998.10	1.90	2,158.38
		Total = RM777,017.48							IRR = 6%

Table 11.3. Comparison Between Conventional Loan, BBA and MM.

Price of house = RM400,000	Customer puts = RM40,000	Financier provide = RM360,000	Monthly rental RM2,000
	APR = 10%	APR = 10%	APR = 6%
	Conventional loan	BBA	MM
Monthly payment	3,159.26	3,159.26	2158.38
Total payment in 30 years	1,137,333.60	1,137,333.60	777,017.48
Total interest/Profit to bank	777,333.60	777,333.60	417,017.48
APR	10%	10%	6%
Balance after 15 years	293,992.43	568,666.38	255,775.84

central bank. Nevertheless, the total payments and loan balances are lowest in the MM among the three financing methods. The mathematical derivation for MM in the Appendix shows that the return to the MM is solely determined by the rental rate, which in this case is 0.5% per month (accordingly the APR is 6%).[13] Interestingly, this return to the financier is neither determined by the initial capital provided by the financier nor the duration of the contract which is usual under debt financing. The return is solely determined by the rental alone as a percentage of the house price. In such a case, financiers of MM would be tempted to finance only homes with high rental rates, whereas it would be in the interest of the customers to negotiate for low rentals.

Some Operational Issues with MM Home Financing

The MM is suited to be practiced, for example, by housing cooperatives where the funds are provided by the members for the benefit of the members themselves. While providing cheaper housing for members, the MM also provides returns to the investing members in the form of rentals and price appreciation. Indeed, observations

[13]The APR in the MM is determined by the rental rate, i.e. the annual rent divided by the original price of the house. In our example, it is (RM2,000 × 12)/400,000 × 100% = 6%. See equation 5 in the Appendix.

show that globally the MM is being successfully practiced in a cooperative setting. Some examples of these are discussed herein before. An additional benefit of implementing the MM by housing cooperatives is that it avoids new money creation as in the fractional reserve banking. By avoiding money creation and operating under a profit-and-loss sharing setting, the MM can bring harmony between the monetary sector and the real economy, check house price increases and thereby likely contribute towards the achievement of the *Maqasid al-Shariah*.

Nonetheless, the MM is not going to be free from operational problems. For example, theoretically the rate of return to MM is determined by the rental rate based on the market rental value, which is very much determined by the location (that affects the property price) and not by market interest rates. As time goes by, the rental value can therefore change, increasing in most cases. It would not then be easy to convince the customer that he now has to pay a higher rental! Keeping track of rentals in many locations can also prove cumbersome.

Some quarters have, therefore, suggested the use of market interest rate, e.g. the London Interbank Offered Rate (LIBOR), Base Lending Rate (BLR), etc. as the benchmark. Adopting this would, of course, render the MM similar to floating-rate conventional financing since the mathematical formulas for the MM turn out to be similar to that of the conventional, as shown in the Appendix (but with interest rate replaced with rental rate). Since many real estate studies have shown the property price as a significant variable determining the rent, some kind of real estate index, like the House Price Index in the Malaysian case[14] can be used as the benchmark. As Shaikh Nizam Yacuby and Muhammad Taqi Usmani mentioned in their 1998 *fatwa*:

> "It is always preferable not to use a benchmark normally used in interest-based transactions, so that an Islamic transaction may not have resemblance with an interest-based transaction."

---

[14]The Malaysian House Price Index is computed and released by the Valuation and Property Services Department, Ministry of Finance, Malaysia. See www.jpph.gov.my.

Hence, unlike the interest rate, a house price index will also be directly linked to the usufruct of the asset.[15] The implementation of MM may also require tax regulations to be amended to allow only rental charges contributing to profit be treated as income to the bank i.e. rental charges contributing to customer's equity must not be taxable (Bilal Bin Farid *et al.*, 2005, p. 16.)

Other practical issues would include issues like what happens if a customer fails to make the rental payments or wants to sell off the property before fully owning it, etc. Also issues about wear and tear, damages to the property due to natural calamities, etc. need to be agreed upon. Such issues can, of course, be incorporated into the MM contract but, nevertheless, are issues to be addressed.

Current Practices of MM-type Home Ownership Schemes

The MM concept has been adopted by a number of Islamic financial services providers worldwide. Successful cooperative-type models include the Islamic Cooperative Housing Corporation (Canada), Ansar Cooperative Housing (Canada) and the Ansar Housing Limited (UK).

The highly successful Islamic Cooperative Housing Corporation (ICHC) based in Toronto, Canada was established in 1981 out of necessity to avoid the Muslim community from engaging in *riba*. It is based on an equity model different from the traditional debt-based mortgage. To join the cooperative, members buy shares in a single equity pool. Once a member accumulates enough shares, the Cooperative (Co-op) buys a house that his family can live in while paying a proportional rent to the Co-op. Thereon, members are required to increase their ownership by investing more money in the Co-op shares. As they do so, the rent goes down in the same proportion until the payments phase out. Eventually, the home buyer surrenders the shares to the Co-op and the Co-op transfers the title. It was reported that under this arrangement, some of their members achieved 100% ownership in seven and a half years as there is flexibility to increase ownership at any time.

---

[15]The index value differs according to locality.

The ICHC model was approved by the renowned scholar Muhammad Taqi Usmani and was also adopted by the Ansar Housing Limited (AHL) of the UK. Other financial institutions that have adopted the *Ijarah* and *Musharakah Mutanaqisah* models include LARIBA in the US, Meezan Bank in Pakistan and Lloyds TSB in the UK.

Integrating MM with FinTech: Crowdfunding a Way Forward

FinTech or financial technology is the use of IT in delivering financial services, that is indeed disrupting the global financial landscape. Very frequently new products and services are being released into the market, particularly by new start-ups. In this part we discuss an innovative method of implementing MM using crowdfunding that is akin to a cooperative model. Crowdfunding is an IT application to source funds from the public globally for some indicated purposes, for example to finance new start-ups, to create a prototype of some new idea, etc. Such ventures may find it difficult to get funding from the banks that are very conservative and risk-averse. Information on the purpose of funds is blasted through the internet globally enticing people to fund those purposes, for some return or otherwise. There is one FinTech company, Ethical and Islamic Crowdfunding (ETHIS) incorporated in Singapore, that crowdfunds the construction of affordable homes in Indonesia. It finances the building of affordable homes for developers who already have ready buyers who have secured financing for the purchase of their homes. This diminishes ETHIS' risk of developer default. The construction component is clearly Islamic and can follow the *Istisna'* contract. The buyer finances the purchase by taking a conventional mortgage or from an Islamic bank. We now discuss and illustrate below our model on how the financing of the purchase too can be done using MM in a crowdfunding framework.

Equity-Type Home Ownership Scheme (ETHOS)

We call the financing of the purchase of a home using MM within a crowdfunding framework, Equity-Type Home Ownership Scheme

(ETHOS). Here, crowdfunders would invest their money with ETHOS that would manage their funds, to finance homebuyers to purchase their homes. The contract used is MM.

Let us say a buyer of ETHIS home would like ETHOS to finance the purchase instead of a conventional bank or an Islamic bank. This buyer could be a crowdfunder (investor) with ETHOS or otherwise. Investors would get returns in the form of dividends from ETHOS, which in turn would get its returns in the form of rentals and capital appreciation from the homes that it finances. Such returns to the investor could be reinvested in order to grow the investors' share in ETHOS. An investor who wants to purchase a home and has at least 10% of the purchase price in his account could indicate the desire to buy to ETHOS. ETHOS would put in the remaining 90% and finance the purchase, using the MM contract, which is a diminishing partnership. The rest would follow as we discussed the contract of MM earlier. ETHOS could also finance a non-investor (non-crowdfunder) to buy a home but the person would have to bring into the partnership at least 10% of the purchase price. Since sustainability is a buzzword nowadays while Shariah-compliant construction is also becoming popular, ETHIS could build homes imbued with sustainable, green features with Islamic designs that fulfill Shariah requirements. We call these homes QAIS homes where Q represents quality, A — affordable, I — Islamic and S — sustainable. ETHIS and ETHOS methods are shown in Figure 11.3 and Figure 11.4 respectively.

ETHOS finances the home-buyer to purchase a home through MM mode. ETHOS is funded through crowdfunding. The buyer can come from among the investors (crowdfunders) themselves or otherwise. If the buyer is among crowdfunders, he has to have accumulated investment of at least 10% of the price of the home he wishes to buy. For outsiders, they have to come up with at least 10% of the price of home upfront. ETHOS would put the remaining 90%, and start-off a MM financing, which is a diminishing partnership based on *Shirkah* and *Ijarah* contracts.

ETHIS MODEL

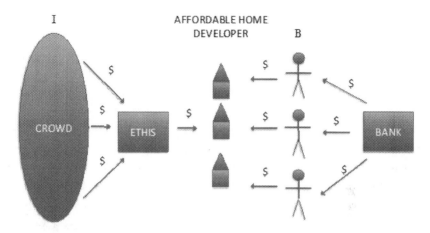

ETHIS finances home developers who have got ready buyers who have secured financing from financial institutions for the purchase of their homes.

Figure 11.3.   ETHIS Finances the Affordable Home Developer.

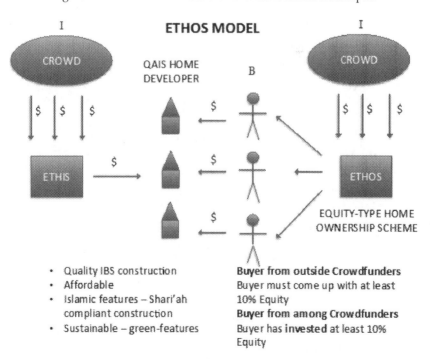

- Quality IBS construction
- Affordable
- Islamic features – Shari'ah compliant construction
- Sustainable – green-features

Buyer from outside Crowdfunders
Buyer must come up with at least 10% Equity
Buyer from among Crowdfunders
Buyer has **invested** at least 10% Equity

Figure 11.4.   ETHOS Finances the Home Buyer.

Conclusion

This chapter looked at the issues of unaffordability of homes due to skyrocketing home prices and the rise in global debt contributed by the mortgage market. We argued this is significantly caused by the structure of the present-day interest-based fiat monetary system. Accordingly, duration of loans has been extended to a maximum period of 35 years and subsequently two-generation loans have also been mooted. The chapter then showed that the present conventional and Islamic contracts additionally pose a number of problems for the home buyers. Interest rates or profit rates are generally high such that the buyers keep paying for the home for a long period of time; and defaults could be very costly particularly with Islamic contracts like the BBA imbued with *bay' al-Inah*. In addition, default is also caused by the structure of the monetary system, brought about by the compounded interest-feature whenever someone in the economy is in default. Accordingly, many home buyers have been bankrupted by this framework. And homes are increasingly becoming unaffordable to most people.

This chapter suggested an equity-type model, as opposed to the debt-type conventional loans and BBA contract, using MM Partnership as the financing mode. While many banks are indeed practicing MM, we argued that the MM would ultimately converge with conventional financing due to arbitrage activities that would make the bank replace rental with interest rates in the formula. Hence, we argued that a Co-op model of MM is most suitable since a Co-op need not be bound by interest rate fluctuations in the economy and comfortably enter into equity-type contracts. Successful cooperative models have been practiced globally since the 1980s. However the cooperative model could also pose a challenge to set up due to local regulations and laws.

Accordingly, we presented in this chapter a FinTech solution, i.e. the use of crowdfunding to mimic an operation like that of a cooperative. Its advantages are as follows:

(1) To enable home-buyers, particularly the younger generation, to own quality homes within an average duration of 10 years.

(2) The financing methods are Shariah-compliant.
(3) Since it is equity-type it shares risk between the buyer and the financier. Hence, even in default the buyer would not be disadvantaged as in the present bank-based financings.
(4) Since crowdfunding makes use of currently existing money, it does not entail new money creation as in the case of bank mortgages where new money (loan) is created by means of fractional reserve banking. Hence it counters skyrocketing home prices.
(5) Helps to alleviate the home unaffordability problem in the country.
(6) Helps to overcome the issue of rising national household debts (with all their socio-economic implications).
(7) Homes are fitted with green-features; supports sustainability initiatives.
(8) Construction of homes are Shariah-compliant and fitted with Islamic features.
(9) Can be implemented where cooperative laws are restrictive.
(10) It also promotes economic justice and stability.

The ETHIS and ETHOS methods together promote the welfare of the people by instituting sustainability and Shariah-compliant features. Anything that protects or promotes these is considered as serving the *maslahah* and hence desirable (Chapra 1992). As al-Ghazzali remarked (Chapra, 1992, p. 1):

> "The very objectives of the Shariah are to promote the welfare of the people, which lies in safeguarding their faith, their life, their intellect, their posterity and their wealth. Whatever ensures the safeguarding of these five serves public interest and is desirable."

As the global finance evolves, being disrupted by FinTech applications with new applications that seem to come into the market frequently, one has to move fast according to the trend. FinTech is the way forward that would level the playing ground for all and promises to bring a lot of good for the global community in the future, in line with the inclusive wealth concept mooted by the United Nations, that promotes just, stable and sustainable systems.

Appendix

P = Price of asset, e.g. a home
B_0 = Financier's contribution into the partnership
C_0 = Customer's contribution into the partnership

Therefore, $P = B_0 + C_0$

R = Periodic rental, eg. monthly
A = Additional periodic payment by customer to redeem the financier's equity faster
$M = R + A$, is therefore, the total periodic payment
Let C_i = the customer's equity (ownership) of the asset in period i

Let the proportion of customer's equity in period i, $r_i = \frac{C_i}{P}$
Therefore,

$$C_0 = C_0$$
$$C_1 = C_0 + r_0 R + A$$
$$C_2 = C_1 + r_1 R + A$$
$$C_3 = C_2 + r_2 R + A$$
$$\vdots$$
$$C_n = C_{n-1} + r_{n-1} R + A$$

Therefore,

$$C_0 = C_0$$
$$C_1 = C_0 + r_0 R + A$$
$$C_2 = C_0 + r_0 R + A + r_1 R + A$$
$$\quad = C_0 + R(r_0 + r_1) + 2A$$
$$C_3 = C_0 + r_0 R + A + r_1 R + A + r_3 R + A$$
$$\quad = C_0 + R(r_0 + r_1 + r_2) + 3A$$
$$\vdots$$
$$C_n = C_0 + R(r_0 + r_1 + r_2 + \cdots + r_{n-1}) + nA$$
$$C_n = C_0 + \frac{R}{P}(C_0 + C_1 + C_2 + \cdots + C_{n-1}) + nA$$

Since $r_i = \frac{C_i}{P}$

Let $x = \frac{R}{P}$, then

$$C_1 = C_0 + xC_0 + A = (1+x)C_0 + A$$
$$C_2 = C_0 + x(C_0 + C_0 + xC_0 + A) + 2A$$
$$= (1 + 2x + x^2)C_0 + (x+2)A$$
$$C_3 = C_0 + x(C_0 + C_0 + xC_0 + A + C_0 + (C_0 + C_0 + xC_0 + A) + 2A)$$
$$+ 3A$$
$$= (1 + 3x + 3x^2 + x^3)C_0 + (x^2 + 3x + 3)A$$
$$\vdots$$

Therefore,

$$C_1 = (1+x)C_0 + A$$
$$C_2 = (1+x)^2 C_0 + (x+2)A$$
$$C_3 = (1+x)^3 C_0 + (x^2 + 3x + 3)A$$
$$C_4 = (1+x)^4 C_0 + (x^3 + 4x^2 + 6x + 4)A$$
$$\vdots$$

$$C_n = (1+x)^n C_0 + \left[\frac{(1+x)^n - 1}{x}\right] A \tag{1}$$

and, of course, the proportion of the customer's equity in the nth period is $r_n = \frac{C_n}{P}$.

Rewriting equation (1), the number of periods taken by the customer to fully own the house is given by, where $C_n = P$,

$$P = (1+x)^n C_0 + \frac{(1+x)^n}{x}A - \frac{1}{x}A$$
$$= (1+x)^n \left[C_0 + \frac{A}{x}\right] - \frac{1}{x}A$$

$$(1+x)^n = \frac{P + \frac{A}{x}}{C_0 + \frac{A}{x}}$$

$$\Rightarrow n = \frac{\ln\left(P + \frac{A}{x}\right) - \ln\left(C_0 + \frac{A}{x}\right)}{\ln(1+x)} \tag{2}$$

Once the rental, R, has been determined and the customer has decided on the period of partnership, i.e. the n, then the periodic amount the customer has to top up additionally is given by

$$A = \frac{x[P - (1+x)^n C_0]}{(1+x)^n - 1} \tag{3}$$

and, the formula for determining the periodic payment is given by

$$
\begin{aligned}
M &= R + A \\
&= \frac{R[(1+x)^n - 1] + x[P - (1+x)^n C_0]}{(1+x)^n - 1} \\
&= \frac{R(1+x)^n - R + xP - x(1+x)^n C_0}{(1+x)^n - 1} \\
&= \frac{x(1+x)^n P - R + R - x(1+x)^n C_0}{(1+x)^n - 1} \quad \text{Since } xP = R \\
&= \frac{x(1+x)^n [P - C_0]}{(1+x)^n - 1} \\
&= \frac{x(1+x)^n B_0}{(1+x)^n - 1}
\end{aligned}
$$

$$\Rightarrow M = \frac{x(1+x)^n B_0}{(1+x)^n - 1} \tag{4}$$

which, interestingly, is similar to the normal annuity formula used for computing the payment in conventional loan calculations.[16] **Hence, mathematically, the normal annuity formula can also be used for *Mushārakah Mutanākisah* calculations, but the periodic interest rate is replaced by the rental rate, $x = \frac{R}{P}$.** Indeed then, the periodic rate of return for *Mushārakah Mutanākisah* partnership is solely determined by the rental rate, $x = \frac{R}{P}$. Therefore, the

$$\text{Internal Rate of Return (IRR) to bank} = \frac{R}{P}. \tag{5}$$

---

[16]Please see footnote 7.

Also,

$$\text{Total payment made to financier} = Mn \qquad (6)$$

$$\text{Total profit to financier} = Mn - B_0. \qquad (7)$$

Note: Since the rate of return (IRR) for the financier is solely determined by the rental rate, $x = \frac{R}{P}$, **irrespective** of the initial capital provided by the financier (B_0) and/or the duration of the partnership (n), the financier may be tempted **to finance only homes with high rental values**; while it is in the interest of the customers to negotiate for low rentals. At the extreme, if the rental is nil, then the *Mushārakah Mutanākisah* financing will become similar to *Qard al-Hassan*.

References

Ahamed Kameel, MM (2002). *The Islamic Gold Dinar*. Subang Jaya, Malaysia: Pelanduk Publications.

Ahamed Kameel, MM (2004). *The Theft of Nations*. Subang Jaya, Malaysia: Pelanduk Publications.

Ahamed Kameel MM and L Moussa (2004). The gold dinar: The next component in Islamic economics, banking and finance. *Review of Islamic Economics*, 8(1), 5–34.

Ahamed Kameel, MM and L Moussa (2005). The Seigniorage of fiat money and the Maqasid al-Shariah: The unattainableness of the Maqasid. To appear in *Humanomics*.

Ansar Finance Limited. http://www.ansarhousing.com [Accessed November 5, 2018].

Bilal Bin Farid, KL Chong, Khairul Fahmi Mahmud, Khairul Hidir Ghazali, Said Anuar Said Ahmad and Syed Hilal Syed Ghazali (2005). Essence of Musharakah Mutanaqisah partnership vs Bai Bithaman Ajil. A project chapter presented for the course *Money, Banking and Capital Markets*, International Islamic University Malaysia, March 2005.

Boualem B and Tariqullah K (1995), *Economics of Diminishing Musharakah*. Jeddah: Islamic Research and Training Institute (IRTI). Research Chapter No. 31.

Chapra, U (1992). *Islam and the Economic Challenge*. United Kingdom: The Islamic Foundation.

"Easy Home" (Diminishing Musharakah). Meezan Bank, Pakistan. www.meezan bank.com [Accessed September 27, 2017].

Global Finance Magazine https://www.gfmag.com/global-data/economic-data/ [Accessed September 27, 2017].

Greco, TH (2001). *Money — Understanding and Creating Alternatives to Legal Tender*, Vermont: Chelsea Green.

Islamic banking grows, with all sorts of rules (May 3, 2005). *The Asian Wall Street Journal*, p. 1.

LARIBA Islamic finance company, United States of America. https://www.lariba.com/sitephp/index.php [Accessed November 5, 2018].

Lietaer, B (2001), *The Future of Money*. New York: Random House.

Lloyds TSB launches Islamic mortgage (March 22, 2005). *Guardian, UK*. https://www.theguardian.com/business/2005/mar/22/islamicfinance.mortgages [Accessed October 4, 2017].

Mbaye, S, M Moreno-Badia and K Chae (2018). Global Debt Database: Methodology and Sources, IMF Working Paper. Washington, DC: International Monetary Fund.

Muhammad Imran Ashraf Usmani (2002). *Meezan Bank's Guide to Islamic Banking*. Karachi: Darul Ishaat.

Muhammad Taqi, U (2002). *Islamic Finance*. Karachi: Maktaba Ma'ariful Qur'an.

Mustafa Omar Mohammed (2005). Islamic financial contracts (Module 2). Fiqh Muamalat Course Material, International Islamic University Malaysia.

Saiful Azhar, R (2005). *Critical Issues on Islamic Banking and Financial Markets*. Bloomington, Indiana: Author House.

Shaikh Nizam Yacuby and Muhammad Taqi Usmani (1998). Fatwa on usage of LIBOR as a benchmark. A Fatwa issued on July 19, 1998.

Chapter 12

Proposed Model on the Provision of Affordable Housing via Collaboration Between *Wakaf-Zakat*-Private Developer

KHAIRUDDIN Abdul Rashid, SHARINA FARIHAH
Hasan and AZILA Ahmad Sarkawi

Introduction

The problem of inaccessibility by the low and middle income Muslim families to decent and affordable housing has been going on for decades but sustainable solutions are yet to be found.

In Islam there exist two key tools to combat poverty namely *wakaf* and *zakat*. However, it is observed that *wakaf* and *zakat* are seldom used to address the problem of inaccessibility to decent and affordable housing among Muslim families. In addition, it is also observed that the *wakaf* and *zakat* institutions seldom work in collaboration especially in addressing the ongoing housing problems facing Muslim families.

Consequently, it is considered apt to find ways as to how *wakaf* and *zakat* can be utilized to solve the problem of inaccessibility by the low and middle income Muslim families to decent and affordable housing.

This chapter is structured into the following parts: Part 1 introduces the chapter. Parts 2 and 3 present a brief review of the problem of inaccessibility by the low and middle income Muslim families to decent and affordable housing and discussions on *wakaf* and *zakat* respectively. Part 4 discusses the development of a proposed model for *wakaf*, *zakat* and private developer to collaborate in addressing the problem of inaccessibility by the low and middle income Muslim families to decent and affordable housing. Part 5 highlights potential issues arising from the collaboration and finally Part 6 provides the chapter's concluding remarks.

The methodology adopted in the study combines extensive review of literature, content analysis and discussions with Shariah experts, *wakaf* and *zakat* administrators and construction practitioners.

The Problem of Accessibility to Decent and Affordable Housing Among Muslim Families

Among the most critical problem currently facing Muslims in Malaysia is the lack of access to decent and affordable housing for the low and middle income families. The problem appears to be more acute among families residing in the cities and big towns due to the higher costs of living. Consequently, many such families have to endure staying in cramped and often unfavorable living conditions.

The terms decent and affordable housing are relative. They mean differently in different contexts or to different authors. However, in this chapter the terms refer to the type of housing that would be appropriate, befitting the lifestyles and incomes of the low and middle income Muslim families. As a benchmark, decent and affordable housing may refer to the Malaysian government's low cost and the *Program Perumahan Rakyat* (PPR) or People's Housing Program.[1]

---

[1]*Program Perumahan Rakyat* (PPR) or People's Housing Program is a government program for the resettlement of squatters and residents who are low income earners. National Housing Department/Ministry of Housing and Local Government is the main implementing agency for the PPR projects throughout Malaysia. PPR consists of two categories, PPR for Rental (PPRS) and PPR for

Access to decent and affordable housing is important to all. Houses not only provide shelter from the sun and rain. More importantly, they serve as a foundation for a conducive and stable family life so that the household can enjoy a decent standard of living — assuming other factors are also conducive — that would lead to a more stable and prosperous society; these being among the key prerequisites to the future wellbeing of the Muslim *ummah*.

> "And Allah has made your houses (that you build) as homes of rest and quiet for you to dwell in; and has made for you out of the skins of cattle tents for dwellings which you find so light (to carry) on the day of wandering and on the day of halting; (also He has made for you) from their wool, fur and hair furnishings and goods for comfort and temporary use (for you) for a time." (Quran 16: 80).

In Malaysia there exist a variety of efforts by various quarters including by the government to alleviate poverty and in the provision of housing for the needy or the low income earners.[2] However, the problem of inaccessibility by the low and middle income Muslim families to decent and affordable housing continues unabated.

Wakaf and Zakat

Wakaf

Wakaf refers to the act of surrendering one's own property so that the same is to be used and for the benefits of others. The key reason for a person to commit the ownership of his property to *wakaf* is to seek the blessings of Allah (*s.w.t*) and to bring oneself near to Him. *Wakaf* may be in the form of land, building, house and cash.

Ownership (PPRM). PPRM houses are sold at prices ranging from RM30,000.00 and RM35,000.00 per unit in Peninsular Malaysia and RM40,500.00 in Sabah and Sarawak. All the houses built under both PPRM and PPRS used the specifications of planning and design of low-cost housing set out in the National Housing Standard for Low Cost Housing Flats (CIS2). For details, see Kementerian Kesejahteraan Bandar, Perumahan dan Kerajaan Tempatan (2017). Program Perumahan Rakyat (PPR). http://ehome.kpkt.gov.my/index.php/pages/view/133 [Accessed August 1, 2017].

[2]Assistance for housing are provided by schemes such as the Amanah Ikhtiar Malaysia or AIM and TEKUN Nasional schemes.

In the context of Malaysia the two commonly discussed *wakaf* are *wakaf am* (general *wakaf*) and *wakaf khas* (specific *wakaf*). *Wakaf* are administered by the respective State Islamic Religious Councils (Majlis Agama Islam Negeri-negeri or MAINs).

"You shall never be truly righteous until you spend of what you love. And whatever you spend is known to Allah." (Quran 3: 92).

Narrated Ibn 'Umar: 'Umar got some property in Khaibar and he came to the Prophet (pbuh) and informed him about it. The Prophet said to him, "if you wish you can give it in charity." So 'Umar gave it in charity (i.e. as an endowment) the yield of which was to be used for the good of the poor, the needy, the kinsmen and the guests. (Sahih Bukhari, Book 55, Hadith 36).

Available literature pointed out that a lot have been done by all the MAINs to administer *wakaf* properties. Some MAINs have even developed *wakaf* land to build hotels and shop-lots which are then rented out to eligible entrepreneurs hence ensuring a steady income stream for the MAINs[3] or formed joint ventures with private companies and property developers to unlock the value of *wakaf* properties and to build low cost housing for eligible Muslim families[4] (Khairuddin, 2017).

However, literature also pointed out the many criticisms concerning the administration of *wakaf* such as (i) while it appears that *wakaf* is rich in assets they are poor in cash, (ii) large parcels of *wakaf* land including those in cities and big towns are left undeveloped and (iii) the lack of expertise to manage the highly valuable land and other assets (MAIAMP, 2014; Khairuddin, 2017).

Consequently, some commentators believe that *wakaf*, as depicted by such a label, and coupled with poor management have led to critiques including that *wakaf* has not been able to realize its potential in terms of poverty alleviation of the *ummah* (Khairuddin, 2017).

[3]Hotel wakaf pertama di Terengganu operasi (April 9, 2013). *Sinar Harian*. http://www.sinarharian.com.my/mobile/hiburan/hotel-wakaf-pertama-di-terengganu-operasi-15-april-1.148262 [Accessed August 1, 2017].
[4]Unlocking land value in Selangor (April 28, 2014). *The Star Online*. https://www.thestar.com.my/business/sme/2014/04/28/unlocking-land-value-in-selangor-agreement-for-development-of-wakaf-sites-signed/ [Accessed August 1, 2017].

Zakat

Zakat or obligatory alms is one of the five pillars in Islam and it is compulsory for all Muslims who fulfill the requirements to pay *zakat* (in the amount and style as stated thereto) and that *zakat* are to be paid to those as prescribed by the Quran:

> "The alms are surely only for the poor and the needy, and for those employed to administer alms, and for those whose hearts have been recently reconciled to the faith, and for captives and those burdened with debts, and (to be spent) for the cause of Allah, and for the wayfarers (stranded on the way). (Such ordinance is) a duty enjoined by Allah. And Allah is All-Knowing, All-Wise." (Quran 9: 60).

In terms of its administration in Malaysia, it is similar to *wakaf* whereby *zakat* is also administered by the MAINs. In the case of *zakat* the various MAINs have been doing very well in terms of collecting and disbursing *zakat* to those that have been identified as eligible recipients (*asnafs*). In addition, the various MAINs have been active in disbursing *zakat* monies not only in the form of cash for daily sustenance but also in the form of other assistance such as scholarships, building or repair of homes and religious schools, education and the like.[5,6]

However, there are criticisms such as the methods of disbursement are too bureaucratic, inefficiencies in collections, whether the disbursements have been effective to assist the needy in the shorter and longer terms (as defined by the Quran) and why is it that while collections have been on the rise and so too the disbursements year after year yet there are still poverty among members of the Muslim *ummah*.[7,8] Surpluses in the amount of *zakat* money (i.e. annual collections higher than annual disbursements) led to *zakat* being rich in cash but lacking in sustainable productivity. As such

---

[5]See for example, Lembaga Zakat Selangor (2013). *Majalah Asnaf, Terbitan 2/2013*. Selangor: Lembaga Zakat Selangor (MAIS).

[6]Lembaga Zakat Selangor (2012). *Laporan Pengurusan Zakat Selangor 2012*. Selangor: Lembaga Zakat Selangor (MAIS), p. 49.

[7]See Lembaga Zakat Selangor (2013), discussions in Asnaf, Terbitan 1/2013, p. 20–22 and 2/2013, p. 8–9; p. 12–13.

[8]See Lembaga Zakat Selangor (2012), p. 24.

zakat, under the current state of affairs, has also been seen as unable to effectively help the *ummah* get out of the poverty trap.

The Key Issues Concerning Affordable Housing, *Wakaf* and *Zakat*

In terms of administration, broadly both *wakaf* and *zakat* are placed under one roof i.e. under the respective MAINs, but in reality, both are administered differently, the reason being *wakaf* and *zakat* have different sets of objectives and recipients as prescribed by the injunctions in the Quran and *hadiths*. In the context of the current chapter, several questions could be asked regarding the administration of *wakaf* and *zakat*, topping the list includes:

- Would it be permissible and more practical for *wakaf* and *zakat*, both governed under the same organ, i.e. the respective MAINs, to share their resources and work together, in a big way, towards the eradication of poverty of the Muslim *ummah*?

- Why can't *wakaf* (that is rich in assets but lacking in cash) collaborate with *zakat* (that is rich in cash but lacking in assets) to work together towards the provision of decent and afford-able housing for the needy and poor Muslim families? That is, why can't *wakaf* allow its land as sites for decent houses and for *zakat* to provide capital and pay for the completed houses and subsequently allocate the units to eligible beneficia-ries of *zakat*? Would not such collaboration ensure that risks in terms of marketability of the completed units be minimized and the plights of the homeless or occupants of indecent housing among the Muslim *ummah* be addressed? The current practice of unlocking the value of *wakaf* land by way of forming joint ventures with property developers is commendable but such a move, especially when it only concerns unlocking the value by way of building and renting out high cost residential and other properties would not, in the long run, help to assist in the provision of decent and affordable housing for the needy Muslim families.

- What could be the hurdles for *wakaf* and *zakat* to collaborate in a truly professional manner?

If the reluctance to collaborate is due to accounting: to ensure properties under *wakaf* are utilized according to the currently available *wakaf* injunctions and the same is also true for *zakat* then surely through prudent and modern management and proper book-keeping, including the application of Information Communication Technology (ICT), the potential mixing of the funds or properties could be alleviated.

If the problem stems from the lack of rulings on *fatwas*, then Muslim scholars must sit down and look into the matter in the most urgent manner. Extensive review of available literature produced no evidence indicating the presence of specific prohibition against such collaboration, nor the presence of *fatwa* — either way — been made on the matter. Consequently, the concept *Al-asl fi al-ashya' al-ibahah* (the presumption of legality) could be applied. The Prophet Muhammad (pbuh) is reported to have said:[9]

> "Reconciling between Muslims is permissible, except reconciliation that forbids something that is allowed, or allows something that is forbidden." (Sunan Ibn Majah, Book 13, Hadith 46)

If the lack of expertise is the reason for such an opportunity to fail to be identified and realized then the various MAINs should consider capacity building, buying in expertise or having them in-house. Such experts may include Quantity Surveyors, Architects, Engineers and Property Managers. The fact that currently almost all of the administrators in the various MAINs are not coming from any of these categories of expertise is a serious concern.

In the context of *wakaf* and *zakat*, apart from meeting the religious requirements, the collected assets and funds should be effectively utilized for the betterment of the *ummah*, hence in the current discussion, a kind of property development set-up within the *wakaf* and *zakat* authorities should be carefully considered.

Consequently, it is considered possible that *wakaf* and *zakat* institutions, both under the ambit of the respective MAINs, could work, and in fact should work together towards the provision of housing for the needy Muslim families.

---

[9]See Azman Mohd Nor and Saidatolakma Mohd Yunus (2012) in Kobayashi, Khairuddin *et al.* (2012), pp. 49–64.

Wakaf-Zakat-Private Developer Joint Venture Model in the Provision of Affordable Housing for Needy Muslim Families

Joint Venture in General

Theories on conventional joint ventures concern broadly with tactical arrangements whereby two or more entities agree to cooperate and bring together their resources and expertise to carry out agreed works, the key objective being for profit and/or mutual benefits. Consequently, costs and risks of a project are spread, there is better access to financial and other resources, access to special expertise or new technologies, pre-empting competition and creation of stronger competitive units (Kobayashi, Khairuddin *et al.*, 2009, p. xxi–xxii).

However, it is worthy to note that while on the one hand there are immense merits to be achieved by way of forming joint ventures there are also pitfalls. According to Kobayashi, Khairuddin *et al.* (2009) the pitfalls include constraints in contractual relationships among the parties. The potential sources of the constraints could be due to factors such as differences in work styles and cultures, institutional problems and uncertainties and risks associated with the project.

Joint Venture from the Islamic Perspective

From the Islamic perspective, collaborations among Muslims either in general or in businesses, are highly encouraged. The verses in the Quran promoting collaborative works or joint ventures include:

> "... Help one another in furthering virtue and God-consciousness, and not in what is wicked and sinful..." (Quran 5: 2).

According to Sheikh Yusuf Al-Qaradawi:[10]

> "It may be said that Allah (Glory be to Him) has distributed talents and wealth among human beings according to a wise plan of apportionment.

---

[10]In his well-known book, Yusuf, Al-Qaradhawi (n.d.). *The Lawful and the Prohibited in Islam*. Al-Falal Foundation. https://thequranblog.files.wordpress.com/2010/06/the-lawful-and-the-prohibited-in-islam.pdf [Accessed May 9, 2018].

We find many a talented and experienced individual who does not possess much wealth or none at all, while others have a great deal of money but little or no talent. Why, therefore, should not the wealthy person turn over to the one possessing talents some of his wealth to invest in a profitable business, so that the two may benefit from one another and share the profits according to some agreed-upon formula? In particular, business ventures on a large scale require the cooperation of many investors. Among the populace we find a large number of people who have savings and excess capital but who lack time or the capability of investing it. Why should not this money be pooled and placed under the management of capable people who will invest it in significant, large-scale projects?

We maintain that the Islamic Shariah did not prohibit cooperation between capital and management, or between capital and labor as these terms are understood in their Islamic legal sense. In fact, the Shariah established a firm and equitable basis for such cooperation: if the owner of capital wishes to become a partner with the working man, he must agree to share all the consequences of this partnership. The Shariah lays down the condition that in such a partnership, which is called *al-mudarabah* or *al-qirad*, the two parties should agree that they will share the profit if there is profit and loss if there is loss in a proportion agreed upon in advance. This proportion can be one-half, one-third, one-fourth, or any other proportion for one party and the remainder for the other party. Thus the partnership between capital and labor is that of two parties with joint responsibility, each having his share, whether of profit or loss, and whether much or little. If, in the balance, the losses exceed the profits, the difference is to be charged against the capital. This arrangement is not surprising, for while the owner of the capital has suffered a loss in his wealth, the working partner has lost his time and effort."

Proposed Joint Venture Model Between *Wakaf*, *Zakat* and Private Developer

There is potential for *wakaf* and *zakat* to collaborate in the provision of appropriate and affordable housing for the low and middle income Muslim families. In an effort to enrich the collaboration, it is considered most appropriate for *wakaf* and *zakat* to collaborate with those with the necessary resources and expertise in housing and property development i.e. the private developers.

Thus, it is thought to be appropriate for a tripartite joint venture to be formed i.e. between *wakaf*, *zakat* and private developer. In the conventional sense, such an endeavor would mean that there is separation between the design, construction and maintenance, and issues on funding, availability of expertise and the running of the completed premises would be fragmented. In addition, risks would be difficult to manage and this would not be appropriate for entities like *wakaf* and *zakat*.

Consequently, a more modern approach in project funding and delivery coupled with a more secured joint venture agreement between the key parties, with risks properly identified and appropriately distributed among the parties considered most able to handle them and deemed appropriate for entities such as *wakaf* and *zakat* is herein proposed. The full idea of the proposal is designed along similar lines to a typical concession contract and to be undertaken on the basis of Design-Build-Finance-Operate-Maintain (DBFOM).

Principally, a joint venture is to be formed on the basis of the *wakaf* and *zakat* authorities becoming the key shareholders: *wakaf* by way of its land being put into the venture as its portion of the share and *zakat* being rich in cash would provide the appropriate capital injection to the project. In an effort to increase the marketability of the venture and minimize project and funding risks to the key partners, it is recommended that these two entities: *wakaf* and *zakat* form a joint venture in the style of a Special Purpose Vehicle (SPV) with a private developer. It would be the SPV that would undertake the project via the DBFOM style of project implementation. The partners to the SPV will sign a joint venture agreement among themselves.

Under the DBFOM arrangement, there will be a concession agreement between *wakaf* as the principal and the SPV as the concession holder. The SPV via a non-recourse funding arrangement will be paid the costs associated with the DBFOM and profit based on agreed terms including carefully crafted and defined Key Performance Indicators (KPIs). Figure 12.1 illustrates the style of the joint venture and the concession agreement.

(1) *Wakaf* authority agrees for its land to be developed for the construction of affordable housing and related infrastructures.

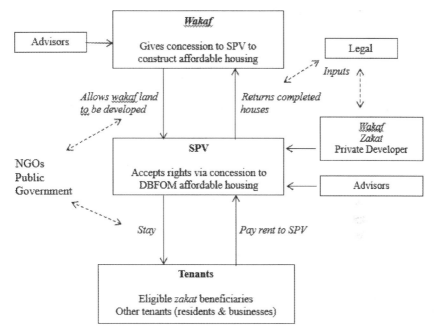

Figure 12.1. Proposed *Wakaf-Zakat*-Real Estate Developer SPV Model for Affordable Housing.

Source: Revised version of Khairuddin, *et al.* (2014).

(2) *Wakaf* and *zakat* authorities plus other key players form an SPV. The joint venture agreement between the parties forming the SPV must recognize the two key partners namely *wakaf* (share in the form of the land) and *zakat* (share in the form of capital injection for the project). The proportion of the housing units between *wakaf* and *zakat* and the level and style of risk and profit sharing should be spelled out in the agreement subject to Shariah's permissibility.

(3) *Wakaf* authority and SPV sign concession agreement. The ownership of the land remains with *wakaf* in perpetuity but allows the SPV to occupy it for the purpose of DBFOM the buildings and infrastructure constructed thereon. The SPV is given permission to collect rents to cover costs for DBFOM plus agreed profit based on agreed KPIs.

(4) Profits would be distributed among the partners in the SPV agreement.
(5) Upon completion of construction, the SPV manages and maintains the building.
(6) The houses and commercial units will be rented out to residents comprising *zakat* beneficiaries, other residents and commercial tenants. The presence of these other tenants would increase the marketability of the venture and potentially increases the income stream of the SPV. The business premises offer conveniences to residents.
(7) *Zakat* beneficiaries stay free, as payment to the SPV is made on their behalf by the *zakat* authority. Selection of tenants would be done by the *zakat* authority based on eligible criteria. Other residents and businesses pay rent to SPV.
(8) Upon expiry of the concession agreement, the land and building are surrendered to the *wakaf* authority. The duration of the concession agreement will be contingent upon factors such as capital injection, maintenance level of services, etc. In short, only benefits (*munafaah*) arising from the land are allowed to be leased out to the tenants.
(9) *Zakat*, thereafter, may want to consider renewing the lease with *wakaf* so as to ensure *zakat* beneficiaries can continue staying or they may want to enter into short-term rental agreements.
(10) *Wakaf*, at the end of the concession period, continues to own the land plus the building sitting on it.

Issues Arising from the Joint Venture, the Concession Agreement and Their Solutions

Issue 1 — Ownership of Wakaf Land

Only land under the category of *wakaf* am should be considered for development. The ownership of *wakaf* land remains with the *wakaf* authority. Under the concession agreement, the SPV is given legal possession to occupy and develop the site for the duration of the agreement only. Upon the conclusion of the concession agreement, the land and the building sitting on it will become the property of *wakaf*. In addition, during the concession period, i.e. after the building has been completed, potential income from the rental of the premises will also become part of the income of *wakaf* as *wakaf* is a shareholder in the SPV.

Issue 2 — Fragmentation of Wakaf Land and Difficulties to Secure a More Appropriate Site

In an effort to ensure marketability of the venture, due to the lack of potentially "good" sites or available land being fragmented in size, it is proposed that land substitution through the concept of *Istibdal*[11] is applied. It is proposed that the *istibdal* is only applied to land belonging to *wakaf am* (general *wakaf*).

Issue 3 — Can Zakat Money Be Used as Capital in the Venture?

In the context of utilizing *zakat* funds for assisting the *asnafs* with their housing needs, the *Lembaga Zakat Selangor* for example has been implementing a series of initiatives such as the construction of houses for *asnafs*, assistance in payments of rent on behalf of *asnafs* and house repairs. Consequently, provision of housing via the *zakat* funds is akin to an extension of the initiatives.

The Quran has provided clear guidance on how *zakat* money should be spent. It is not the intention of the authors to meddle with what has been practiced thus far. However, in examining in detail the amounts of *zakat* collected and *zakat* disbursed in a given year, there seems to be surpluses, i.e. collection is more than the amount disbursed and the situation happens quite frequently.

In addition, there are scholars categorizing *zakat* into immediate and progressive. Immediate *zakat* is the kind of *zakat* that must be disbursed immediately. However, progressive *zakat* is an amount of *zakat* money that could be used for the purchase of tools for use in an employment that subsequently would enable the beneficiaries to earn a living.

---

[11]Discussion on *Istibdal* is not within the scope of this chapter.

It is proposed that, after careful study and detailed consideration made, only the amount in surplus could be used in the affordable housing venture and that such use is categorized under the category of progressive *zakat*. Consequently, beneficiaries of *zakat* only, to be determined by the *zakat* authority, can be allocated units in the completed housing scheme under the allocation accorded to the *zakat* authority as defined in the joint venture agreement. Further and longer-term ownership issues surrounding such *zakat* beneficiaries would have to be resolved.

Issue 4 — The Potential of Zakat Experiencing Shortage of Cash and Being in Need of Cashing in Their Shares in the Joint Venture

In the unlikely event[12] that *zakat* requires cashing in their shares in the joint venture then a provision of "buy-back" should be incorporated into the joint venture agreement. Parties that are able to take over the shares allocated to *zakat* should be identified in advance and a provision to that effect should be among the terms and agreement of the contract. The possibility of cashing in by *zakat* could not be ignored because an event might occur leading to the *zakat* collection becoming negative *vis-à-vis* disbursement and an amount is urgently required to ensure that payment for immediate *zakat* would not be delayed or compromised. Alternatively, a Shariah-compliant bond should be secured prior to the go-ahead.

Issue 5 — Sustainability of the Project Beyond the Concession Period

The project should have the longer-term issue of sustainability built in such as the appropriate infrastructure and the presence of appropriate commercial units and the like. Such steps must be taken so that not only will the project unlock the real value of the

---

[12]Unlikely because prior to the venture being given the go-ahead the relevant authorities must conduct an in-depth feasibility study and due diligence.

otherwise idle *wakaf* land and provide accommodation to the needy Muslim families and beneficiaries of *zakat* but the project should be viable in the longer term for the benefit of the Muslim *ummah*.

Issue 6 — The Problem of Lack of Expertise in MAINs

The various MAINs are entrusted with assets worth billions with the value of some of the assets worth much more if efforts to unlock their values and potential are undertaken in the most appropriate and professional manner.

MAINs should consider either importing the services of experts such as Quantity Surveyors, Property Managers, Facility Managers, Architects, Engineers, Lawyers and Accountants or to have them employed as in-house experts.

Issue 7 — Allocation of Affordable Housing Units to Eligible Zakat Beneficiaries

On the completed and available units in the proposed model, the units allocated under the *wakaf* portion would not be much of an issue as the units would be leased or rented out at a rate appropriate for low income households.

However, for the units allocated under the *zakat* portion, a systematic selection and allocation procedure must be designed and put in place. The system must ensure the target group would not be overlooked: transparency, accountability and fairness to all shall prevail.

Issue 8 — Delays and Abandoned Project

Among the many issues concerning property development are delays and abandoned projects. Much has been written about this and a variety of systems have been put in place to combat the problem. In short, the authority entrusted to undertake the project must ensure that strict rules and regulations must be followed and that defaulters must be appropriately and severely punished.

Conclusion

This chapter identified issues related to access to decent and afford-able housing among Muslim families. It discussed the potential of *wakaf*, *zakat* and private developer to collaborate in the provision of affordable housing and proposed an appropriate model. Admit-tedly, the model proposed herein is still in its infancy. In addition, the study has been carried out within the context of its inherent limitations: methodology, skills and knowledge to understand *fiqh* and the lack of available and published materials. Thus, there would be issues and challenges, both from the practice and Shariah perspectives that require further studies, deliberation and fine-tuning. The authors welcome constructive comments and criticisms.

Acknowledgment

The study reported herein is funded by the Transdiciplinary Research Grant Scheme, Ministry of Higher Education, Malaysia. Program Title: Development of a Protocol to Empower *Wakaf-Zakat* in the Provision of Housing for the *Ummah*. Project Title: A Study into Procurement and Project Delivery System to Empower *Wakaf-Zakat* in The Provision of Housing for the *Ummah*. Project ID: TRGS16-01-001-0001.

References

Azman, MN and MY Saidatolakma (2012). The application of BOT contract as a mode of financing in developing *waqf* land: Malaysia experience, In *Joint Ventures in Construction 2*, K Kobayashi *et al.*, pp. 49–66. London: Thomas Telford.

Hotel wakaf pertama di Terengganu operasi (April 9, 2013). *Sinar Harian*. http://www.sinarharian.com.my/mobile/hiburan/hotel-wakaf-pertama-di-terengganu-operasi-15-april-1.148262 [Accessed August 1, 2017].

Kementerian Kesejahteraan Bandar, Perumahan dan Kerajaan Tempatan (2017). Program Perumahan Rakyat (PPR). http://ehome.kpkt.gov.my/index.php/pages/view/133 [Accessed August 1, 2017].

Khairuddin, AR, H Sharina Farihah and AS Azila (2014). Development of a proposed waqaf-zakat-real estate developer joint venture model to address the issue of affordable housing for the low and middle income Muslim families. In *Proc. World Conference on Islamic Thought and Civilization (The Rise and Fall Of Civilization: Contemporary States of Muslim Affairs)*. Ipoh, Perak: August 18–19, 2014.

Khairuddin, AR (2017). Sumbangan sektor hartanah wakaf kepada penyelesaian isu perumahan negara. Presentation at The Seminar Wakaf FSPU UiTM.Shah Alam, Selangor, Malaysia: September 20, 2017.

Kobayashi, K, AR Khairuddin, G Ofori and S Ogunlana (eds.) (2009). *Joint Ventures in Construction*. London: Thomas Telford.

Lembaga Zakat Selangor (2013). *Majalah Asnaf, Terbitan 2/2013*. Selangor: Lembaga Zakat Selangor (MAIS).

Lembaga Zakat Selangor (2012). *Laporan Pengurusan Zakat Selangor 2012*. Selangor: Lembaga Zakat Selangor (MAIS).

Majlis Agama Islam dan Adat Melayu Perak (MAIAMP) (2014). In Khairuddin, AR, H Sharina Farihah and AS Azila (2014). Development of a proposed waqaf-zakat-real estate developer joint venture model to address the issue of affordable housing for the low and middle income Muslim families. In *Proc. World Conference on Islamic Thought and Civilization (The Rise and Fall Of Civilization: Contemporary States of Muslim Affairs)*. Ipoh, Perak: August 18–19, 2014.

Unlocking land value in Selangor (April 28, 2014). *The Star Online*. https://www.thestar.com.my/business/sme/2014/04/28/unlocking-land-value-in-selangor-agreement-for-development-of-wakaf-sites-signed/ [Accessed August 1, 2017].

Yusuf, Al-Qaradhawi (n.d.). *The Lawful and the Prohibited in Islam*. Al-Falal Foundation. https://thequranblog.files.wordpress.com/2010/06/the-lawful-and-the-prohibited-in-islam.pdf [Accessed May 9, 2018].

Chapter 13

Shariah Compliance Risk-Sharing in Islamic Contracts

Masamitsu ONISHI and Kiyoshi KOBAYASHI

Introduction

Shariah prohibits *riba*, *maysir* and *gharar*. This chapter focuses on the concept of *gharar* which is literally directed as uncertainty or risk in common. Shariah prohibits transactions which contain the concept of *gharar*. However, this does not mean that Shariah prohibits any transaction which is associated with uncertainty. Construction projects essentially entail a considerable degree of uncertainty. The fact that a number of construction projects are implemented by mobilizing Islamic finance implies that *gharar* does not simply mean the concept of risk termed in the conventional society. Prohibition of *gharar* is rather applied to a specific type of transaction and contract that entails uncertainty.

Construction projects that mobilize Islamic finance requires their contracts to conform to Shariah. Conventional Islamic jurisprudence derives rules of Shariah based on the interpretation of holy Quran and *hadith*. However, there is no universal consensus on the rigorous definition of *gharar*. In such a circumstance, a mathematical approach to define the concept of *gharar* has emerged (Al-Suwailem, 2000, 2006). This approach plays a great role in defining the concept of *gharar* in a rigorous way. However, as will be mentioned, the existing approach is not complete in the sense that it is not effective in judging Shariah compliance for cases in construction contracting. This study theoretically investigates conditions for Shariah incompliance by modelling the risk-sharing problem in complete contract setting by employing game theory.

This chapter aims at proposing a theoretical hypothesis concerning Shariah compliance in a construction contract under uncertainty. In practice, the judgment on Shariah compliance must be assessed by a Shariah Board. Nevertheless, posing a theoretical hypothesis and falsification should contribute to our deep understanding about Shariah compliance in construction contracting.

The Basic Idea about *Gharar*

Muslims are required to comply with the Shariah. Shariah consists of the holy Quran and *hadith*. Quran and *hadith* provide the obligations in life and forbidden rules. However, Quran and *hadith* do not provide the substantial principle or normative concept in Shariah. Therefore, after the death of Prophet Muhammad (pbuh), rules of Shariah for new cases have been derived from interpretation of the Quran and the *hadith*. Islamic jurisprudence is the body of knowledge concerning the rules derived from the Quran and the *hadith*.

In Islam, transactions including *riba*, *maysir* and *gharar* are prohibited. When the capital for a project is procured from an Islamic financial institution, a contract must be checked to confirm that it does not include these prohibited items.

Table 13.1.   Poles of Interpretation of *Gharar*.

	Scope of permissibility	What to avoid?	Supporters (schools of law)
Position A	Restrictive	Ignorant, non-existence	*Shafi'i, Hanafi*
Position B	Broad	Enmity, distraction from prayer	*Hanbali, Maliki*

Source: Vogel and Hayes, 1998.

In the case of *gharar*, its position may be interpreted as per Position A or Position B as indicated in Table 13.1.

Position A, as a general rule, voids sale of nonexistent or uncertain objects without any consideration of degree of risk involved. The *Shafi'i* and *Hanafi* schools are most characterized by position A that leads to highly restrictive rules. For example the *Shafi'i* school prohibits sales of absent specific objects altogether. Both schools prevent sales of objects present but invisible (such as carrots underground). The *Hanbali* and *Maliki* schools move slightly toward Position B. They admit for example the sale of absent objects by a complete description. In such cases the contracts are binding as long as the subject matters agree with the description. The *Malikis*, though not some *Hanbalis*, allow the sale of present but unviewed items such as underground carrots or nuts in the shell. The *Maliki* school often distinguishes itself by allowing sales that other schools prohibit, when circumstances indicate that *gharar* will be mild (Vogel and Hayes, 1998).

Position A classifies examples of *gharar* referred in *hadith* into the following three categories:

- Ignorant about goods
- Nonexistence of goods
- Goods which are not under the administration of the seller

The third case can be regarded as a special case of the second, i.e. nonexistence of goods. Position A has two requirements for an effective contract: i) sufficient information and knowledge concerning the transaction and ii) existence of transacted goods. Compensating for the absence of these requirements by adjusting the transacted price is not allowed (Vogel and Hayes, 1998).

Examples of the case of ignorance about goods are *munābadha* and *mulāmasa*. *Munābadha* is a manner of transaction where the seller throws the goods to the buyer and does not give an opportunity to see it carefully. *Mulāmasa* is a manner of transaction where the seller regards the buyer as accepting to purchase the goods once the buyer touches it. Examples of the case of nonexistence of goods are *habl habla* and *hasat*. *Habl habla* is sale of goods which does not exist such as sale of a baby camel in the pregnant mother's body or grandchild camel. *Hasat* is sale of goods which cannot be specified clearly.

Position B is based on a typical example of *gharar*. Imam Malik ibn Anas — the founder of Maliki school of law — states in *muwatta'* (Bewley, 2001): "Included in *gharar* and risky transactions is the case in which a man whose camel is lost, or his slave has escaped, the price of which is (say) 50 dinar, so he would be told by another man: I will buy it for 20 dinars. Thus if the buyer finds it, the seller loses 30 dinars; if not, the buyer loses 20 dinars."

Again here we are reminded of Ibn Taymiyya's clear explanation about this case from Al-Suwailem (2000):

> "*Gharar* describes things with unknown fate. Selling such things is *maysir* and gambling. This is because when a slave runs away, or a camel or a horse is lost, his owner would sell it conditional on risk, so the buyer pays much less than its worth. If he gets it, the seller would complain: you have 'gambled' me and got the goods with a low price. If not, the buyer would complain: you've 'gambled' me and got the price I paid for nothing. This will lead to the undesired consequences of *maysir*, which is hatred and enmity, besides getting something for nothing, which is a sort of injustice. So *gharar* exchange implies injustice, enmity and hatred."

Assessing *Gharar* with Regret Theory

Existing Mathematical Approach for Defining Gharar

Al-Suwailem (2000, 2006) tries the mathematical approach to define the concept of *gharar* rigorously. He argues that *gharar* is risk associated with transactions of zero-sum structure based on the idea of *maysir*. He applies regret theory (Loomes and Sugden, 1982) to analyze the transaction of "lost camel" mentioned before. In the transaction of lost camel, the absolute values of regret of players are the same but its signs are opposite. This is a feature of payoff structure regarding the zero-sum game. In a zero-sum game, if one gains benefit, one suffers loss. Therefore, it is argued that the zero-sum structure should be prohibited since win-lose relationship caused by zero-sum provokes the feeling of enmity between the parties to a contract. In addition, in the non-zero-sum game where the sum of players is zero, the contract is deemed as *gharar* if win-lose relationship is dominant. On the contrary, if win-lose relationship is not dominant, that contract is not *gharar*. The above discussion will be discussed in detail in the next subsection.

The transaction of lost camel analyzed by Al-Suwailem (2000) has a unique payoff structure in that the seller can gain the camel even though a contract is not concluded. It implicitly assumes that the seller has an outside option to sell the camel to a third person even though the seller does not conclude the contract with the specified buyer. Regret theory is employed in order to understand this unique game structure. However, Al-Suwailem (2000) only makes an argument on the payoff structure of transactions and leaves the issue of decision-making to parties. Because of this limitation, the model cannot apply to the analysis for transactions with non-zero-sum structure. This chapter claims that it is possible to derive an equilibrium solution of transaction of lost camel by re-defining the payoff function. Furthermore, we will analyze the conformity of transactions with non-zero-sum structure with Shariah by applying an extensive form game model.

Differently from Al-Suwailem (2000), El-Gamal (2001, 2006) argues that contracting entailing uncertainty is permissible as long as it contributes to improving the efficiency in the sense of Pareto. El-Gamal employs prospect theory by assuming people's rationality in decision-making. El-Gamal argues that people with bounded rationality make decisions according to prospect theory. If people make decisions according to prospect theory, it is possible that inefficient transactions are realized. If a more efficient contract exists, such a contract is *gharar*. The focal point of Al-Suwailem's argument concerns the principle of equity regarding the ex-post payoff structure, whereas that of El-Gamal's argument concerns the principle of efficiency at the pre-contractual timing. Our model explained later can deal with both the principles of ex-post equity and ex-ante efficiency criteria.

Despite some limitations of Al-Suwailem's model, his idea must be appreciated in terms of his contribution to developing a framework to define the concept of *gharar* rigorously. Among others, his discussion regarding regret win-lose relationship of payoff structure provides a fundamental aspect to argue the issue of equity. The remaining part of this section introduces his analytical framework in detail and identifies inherent problems in his discussion.

Regret Theory

Expected utility theory has been developed as a normative decision-making theory under uncertainty thanks to the pioneering contribution by Neuman and Morgensterun (1944). On the other hand, the expected utility theory has been generalized responding to some critiques regarding contradiction with the reality claimed by such as Allais (1953), Ellsberg (1961) and Savage (1972). Regret theory has been proposed in the process of such a generalization process of expected utility theory. Let x_i^s, x_j^s denote the result when a player chooses two possible alternatives θ_i, θ_j at the state of nature $s(s = 1, \ldots S)$. The state of nature is stochastic variable at the time when the player makes decisions. Regret theory defines a revised utility function which depends on x_i^s, x_j^s in order to overcome the difficulty in expected utility theory pointed out by Kahneman and Tversky (1979), and Loomes and Sugden (1982).

The revised utility function called Regret and Rejoice function (hereinafter called RR function) when a player chooses θ_i, not θ_j at the realized state of nature s is defined as

$$u_{ij}^s = R(x_i^s - x_j^s) \tag{1}$$

where $R(\cdot)$ is non-decreasing function in $x_i^s - x_j^s$ and $R(0) = 0$. $R(x_i^s - x_j^s) < 0$ implies that the player feels regret by choosing θ_i rather than θ_j, whereas $R(x_i^s - x_j^s) > 0$ implies that the player feels rejoice. Given the probability function $p(s)$, the expected value of revised utility E_i is formulated as

$$E_i = \sum_{s=1}^{S} p(s)u_{ij}^s. \tag{2}$$

Table 13.2. Payoff Transaction of Lost Camel.

| | The seller | | The buyer | |
	Camel found	Camel not found	Camel found	Camel not found
Accept the Contract	20	20	$-20 + 50 = 30$	-20
Reject the Contract	50	0	0	0

Table 13.3. Regret Payoff.

| | The seller | | The buyer | |
	Camel found	Camel not found	Camel found	Camel not found
Accept the Contract	$20 - 50 = -30$	$20 - 0 = +20$	$-20 + 50 = +30$	$-20 - 0 = -20$
Reject the Contract	$50 - 20 = 30$	$0 - 20 = -20$	$0 - 30 = -30$	$0 - (-20) = 20$

The rational choice of θ_i, θ_j is represented as

$$\max\{E_i, E_j\}. \tag{3}$$

Transaction of Lost Camel

Al-Suwailem (2006) applies regret theory to transaction of lost camel referred in *Muwatta* to characterize transactions including *gharar*. Following Al-Suwailem's (2006) analysis, we assume the linear RR function of player $k(k = \alpha, \beta)$ as

$$R^k(x_i^{sk} - x_j^{sk}) = x_i^{sk} - x_j^{sk} \tag{4}$$

where x_i^{sk} denotes the payoff of player k when he chooses the alternative $\theta_i(i = 1, 2)$. The seller $(k = \alpha)$ sells the right to obtain the lost camel to the buyer $(k = \beta)$. At the time when the seller sells the right to obtain to the buyer, it is not certain whether the lost camel will be found or not. There are two possible states: (1) the camel is found $(s = 1)$ or (2) the camel is not found $(s = 2)$. There are two alternatives for player k: (1) θ_1^k: accepting the contract or (2) θ_2^k: rejecting the contract. Given that the cost of seeking the lost camel is negligible, the state-contingent payoff is written as Table 13.2. From the state-contingent payoff, the value of RR function (hereinafter called RR payoff) depending on the choice of alternatives for each player can be derived as Table 13.3. The RR payoff satisfies

$$R^\alpha(x_1^{s\alpha} - x_2^{s\alpha}) = -R^\beta(x_1^{s\beta} - x_2^{s\beta}) \tag{5a}$$

$$R^\alpha(x_2^{s\alpha} - x_1^{s\alpha}) = -R^\beta(x_2^{s\beta} - x_1^{s\beta}) \tag{5b}$$

which implies the sum of two players' payoff equals zero.

Now consider that the subjective belief of the seller and the buyer regarding the possibility of finding the lost camel is denoted by p and q respectively. Because this possibility is subjective, $p = q$ is not necessarily satisfied. Then the expected revised payoff of the seller $(k = \alpha)$ for each alternative $\theta_i(i = 1, 2)$ is written as

$$E_1^\alpha = -30p + 20(1 - p) \tag{6a}$$

$$E_2^\alpha = 0 \tag{6b}$$

and that of the buyer is written as

$$E_1^\beta = 30q - 20(1 - q) \tag{7a}$$

$$E_2^\beta = 0. \tag{7b}$$

The condition that both the seller and the buyer agree on the contract is $E_1^\alpha > E_2^\alpha$ and $E_1^\beta > E_2^\beta$ and hence

$$\text{The seller: } p < 0.4 \tag{8a}$$

$$\text{The buyer: } q > 0.4. \tag{8b}$$

If the small amount of transaction cost is necessary, the inequality condition is satisfied strongly. (8a) and (8b) is written as $q > 0.4 > p$ which implies the buyer evaluates the possibility of finding the lost camel higher than the seller.

Table 13.4 shows the RR payoff for each possible state. It is easily understood from Table 13.4 that the sum of two players' RR payoff

Table 13.4.   State-Contingent RR Payoff.

	Camel found		Camel not found	
	The seller	The buyer	The seller	The buyer
Accept the contract	−30	30	20	−20
Reject the contract	30	−30	−20	20

is zero for any possible state. The transaction of lost camel has the payoff structure where the sum of RR payoff for any state is zero. We call such a payoff structure as regret zero-sum structure. Note that the regret zero-sum structure holds only when the RR function is linear and $R(0) = 0$. Transaction with the regret zero-sum structure would cause the feeling of enmity since when one player is rejoicing, the other feels regret in any possible state.

Al-Suwailem (2000) argues the feeling of enmity consists of *gharar* referred to as Position B in Table 13.1. In addition, the fact that the contract is concluded implies the seller and the buyer have inconsistent beliefs on the possibility of finding the lost camel. The inconsistency of belief can be interpreted as asymmetric knowledge or information between the players referred to as Position A in Table 13.1.

It is noteworthy that the transaction of lost camel has a unique payoff structure in that the seller gains 50 if the contract is not concluded and the camel is found. This means that the seller has the outside option to sell the found camel to a third person even if he cannot agree on the contract with the specified person. If we assume that the seller does not have the outside option, the payoff of lost camel transaction does not have regret zero-sum structure. Al-Suwailem's approach does not provide a clear answer for the conformity with Shariah.

Regret Win-Lose Condition

We generalize the approach of Al-Suwailem (2006) based on regret theory. Assume that two players, player α and player β make decisions whether they accept a specified contract or not. Player

$k(k = \alpha, \beta)$ chooses either (1) θ_1^k: accepting the contract or (2) θ_2^k: rejecting a contract. x_1^{sk} denotes the payoff of player k at state s when the contract is concluded and x_2^{sk} denotes the payoff of player k at state s when the contract is not concluded. Let the revised utility of player k at state s which realizes after the contracting decision is made as

$$u_1^{sk} = R^k(x_1^{sk} - x_2^{sk}). \tag{9}$$

$$(k = \alpha, \beta; s = 1, \ldots, S)$$

When the followings relations are satisfied, we call "regret win-lose condition holds for state s."

$$R^\alpha(x_1^{s\alpha} - x_2^{s\alpha}) > 0 \quad R^\beta(x_1^{s\beta} - x_2^{s\beta}) < 0 \tag{10a}$$

$$R^\alpha(x_1^{s\alpha} - x_2^{s\alpha}) < 0 \quad R^\beta(x_1^{s\beta} - x_2^{s\beta}) > 0 \tag{10b}$$

If the regret win-lose condition holds, such a transaction may invoke the feeling of enmity which is referred to as a reason of *gharar* by Al-Suwailem (2006).

Win-Win/Lose-Lose Condition

Strategic process of contracting with an outside option, i.e. an option not to take part in a contract, can be formulated as a game theoretical model where a player gains zero payoff by choosing an outside option. Therefore, it is possible to analyze the win-lose relation in the payoff structure without employing regret theory. Al-Suwailem's contribution is to point out that the win-lose condition could lead to the feeling of enmity which should be avoided for the realization of social justice in Islamic society.

If the payoff of a transaction has zero-sum structure, any feasible results are a win-lose relation. However, if the payoff of a transaction does not have zero-sum structure, win-win condition and lose-lose condition may realize. Win-win condition guarantees that both players gain positive payoff when the contract brings benefit, whereas lose-lose condition guarantees that both gain negative payoff when the contract brings loss. Prohibition of *gharar* requires designing a contract to prevent the occurrence of win-lose condition ex-post. Ideal transactions should realize when all parties to a contract gain positive payoff by accepting the contract. The occurrence of win-lose relations contradicts with the ideals of Islam.

Al-Suwailem's approach is limited to the reference to win-lose condition. However, it does not clearly mention the treatment of win-win or lose-lose conditions regarding transactions with the non-zero-sum structure.

Model for Shariah Compliance Assessment

Settings

Contract is a voluntary agreement concerning exchange of goods or wealth between or among more than one entity. Any individual has a right to determine whether he or she makes an agreement or not, which guarantees any entity that enters a contract enjoy more favorable conditions. Voluntary exchange based on the individual's will is encouraged not only in Adam Smith's market fundamentalism but also in Islamic society. *Gharar* is interpreted as a rule to constrain some specific classes of contracting which are possible to be voluntarily agreed on. The normative principle of that constraint implies the possibility that a voluntary agreement does not always guarantee the improvement of all contracting parties.

This study proposes a formal model of project contracting under uncertainty based on a simple extensive form of game with two players. We assume that a contract is complete in the sense that all possible states in the future are describable. Under a complete contract setting, players can describe allocations of welfare for any possible states in the future in advance in a contract. A project is expected to bring positive welfare. Here, payoff of a project is defined as net value of a project, i.e. benefit of the project minus cost. The result of a project is uncertain. Hence, the payoff of a project could be negative. The allocation of project payoff for each possible state is exogenously given. We formalize the problem of Shariah compliance assessment as the problem of assessing the conformity of the payoff allocation rule with non-*gharar* or non-*maysir*. Note that Shariah compliance of project contracting must be assessed not only from the aspect of *gharar* and *maysir* but also from the aspect of *riba*. Therefore, the fact that a project contract does not contain *gharar* does not guarantee its Shariah compliance. Hereinafter, we propose two conditions for guaranteeing non-*gharar* of project contracting: (1) the ex-post equity condition and (2) the ex-ante efficiency condition.

Contracting Game Under Uncertainty

Figure 13.1 shows an extensive form of game which describe the strategic environment of reaching contracting agreement between two players α and β. Players' payoff depends on:

- State of agreement/disagreement on a contract

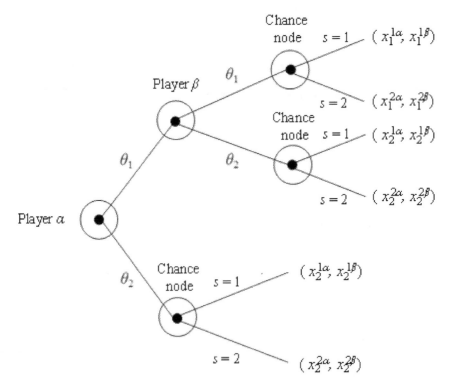

Figure 13.1.   Contract Game Under Uncertainty.

- Risk events uncertain at the time of contractual agreement
- Agreement on the allocation rule of project payoff (only if a contract is concluded).

Let θ_i $(i = 1, 2)$ denote two alternatives for players in the game of contract bargaining, where θ_1 means "accepting the contract" and θ_2 means "rejecting the contract." The contract is concluded only if both of two players choose θ_1. Let δ_i $(i = 1, 2)$ denote the state of contract agreement/disagreement, where δ_1 means the state both players agree on the contract, and δ_2 means the state the contract is not concluded since either or both players reject the contract. The possible risky states which are describable is denoted by $s \in \{1, \ldots, S\} = \Omega$. Let x_i^{sk} denote payoff of player k $(k = \alpha, \beta)$ at the state δ_i $(i = 1, 2)$. Let w_i^s denote project payoff which is the available wealth for two players the state δ_i $(i = 1, 2)$, that is,

$$w_i^s = x_i^{s\alpha} + x_i^{s\beta}. \tag{11}$$

As mentioned above, Al-Suwailem's model assumes players have an outside option to reject a contract. In this case, players gain $x_2^{s\alpha}$ which coincides with the payoff of initial state.

If the contract is concluded, project payoff is allocated between the players. Contractual type can be defined by the method on how to allocate project payoff for any state s between the players. The allocation rule of project payoff is defined by mapping τ from the state space Ω to the real number space R which represents payoff of player α,

$$\tau : \Omega \to R. \tag{12}$$

The mapping τ represents the rule for allocating project payoff and that is the subject of *gharar* assessment. Hereinafter, we call mapping τ as contract type τ. In addition, let τ^s be payoff of player α. Then under the contractual type τ, payoff of player k is represented as

$$x_1^{s\alpha}(\tau) = \tau^s \tag{13a}$$

$$x_1^{s\beta}(\tau) = w_1^s - \tau^s. \tag{13b}$$

(13a) and (13b) means

$$x_1^{s\alpha}(\tau) + x_1^{s\beta}(\tau) = w_1^s, \qquad (14)$$

which implies that the players split the pie of the size of w_1^s.

Now let the players' payoff of contract game be standardized so that the initial payoff of players x_2^{sk} is zero. We define $z_1^{sk}(\tau)$ to be net payoff of player k at state s, i.e.,

$$z_1^{sk}(\tau) = x_1^{sk}(\tau) - x_2^{sk} \qquad (15)$$

and payoff of the initial state

$$z_2^{sk}(\tau) = 0. \qquad (16)$$

Let us define the state-dependent net payoff as

$$w^s = w_1^s - w_2^s. \qquad (17)$$

Given the game with the standardized payoff, two players play the game according to the following sequence of decision-making. (1) Player α determines if he accepts or rejects the contract with the payoff allocation τ. If player α rejects the contract, it is not concluded. (2) If player α accepts the contract, player β determines if she accepts or rejects the contract. (3) If player β rejects the contract, it is not concluded. (4) If player β accepts the contract, it is concluded. The payoff is determined depending on the realization of contract conclusion and the state of nature if the contract is concluded. This extensive form of the game is called contract game Γ.

Sub-Game Perfect Equilibrium

Let us derive a sub-game perfect equilibrium of game Γ. Players have subjective belief concerns about the probability of the occurrence of the risk event. Subjective belief of player k ($k = \alpha, \beta$) at the state s ($s = 1, \ldots, S$) is denoted by $p_k(s)$, where

$$\sum_{s=1}^{S} p_k(s) = 1. \qquad (18)$$

Subjective belief is formed based on the knowledge that the players privately own. The expected payoff of player β at the node of player β is calculated as

$$\begin{cases} \sum_{s=1}^{S} p_\beta(s) z_1^{s\beta}(\tau) & \text{if accepted} \\ 0 & \text{if rejected} \end{cases} \quad (19)$$

Player β will accept the contract only if

$$\sum_{s=1}^{S} p_\beta(s) z_1^{s\beta}(\tau) > 0 \quad (20)$$

is satisfied. If the existence of a small transaction cost for contracting is considered, the strong inequality of formula (20) is the condition of accepting the contract for player β. The following discussion assumes that a small transaction cost for contracting exists.

Then let us analyze the choice of player α. If player β does not accept the contract, i.e., formula (20) is not satisfied, player α gains the payoff $z_2^{s\alpha}(\tau) = 0$ regardless of his decision. Therefore, without loss of generality, player α is assumed to reject the contract if player β is expected to reject it. On the other hand, if formula (20) is satisfied, the contract is concluded upon the acceptance by player α. The expected payoff of player α is derived as

$$\begin{cases} \sum_{s=1}^{S} p_\alpha(s) z_1^{s\alpha}(\tau) & \text{if accepted} \\ 0 & \text{if rejected} \end{cases} \quad (21)$$

Then player α will accept the contract only if

$$\sum_{s=1}^{S} p_\alpha(s) z_1^{s\alpha}(\tau) > 0 \quad (22)$$

is satisfied.

Summarizing the above analytical result, the contract is concluded if both of formula (20) and (22) are satisfied and is not otherwise. The following analysis assumes that common knowledge is shared between the two players, hence $p_\alpha(s) = p_\beta(s) = p(s)$ is satisfied. The case of $p_\alpha(s) \neq p_\beta(s)$ will be discussed subsequently.

Lost Camel Revisited

Under the complete contract setting, parties to a contract can determine the allocation of payoff contingent to the describable state of nature in the future (Salanie, 1997). This study aims to propose a model to assess the conformity of a contract with Shariah focusing on the payoff allocation rule of a complete contract. As discussed earlier, Al-Suwailem proposes a criterion of *gharar* checking if the contingent allocation of payoff satisfies the win-lose condition or not. The zero-sum condition is automatically satisfied in the case of zero-sum game. However, the win-lose condition even for a non-zero-sum game case can be derived by using the contract game proposed here. Let us re-examine the "lost camel" transaction based on the contract game. The payoff of each player if the contract is concluded is

$$\begin{cases} z_1^{1\alpha} = -30 \\ z_1^{1\beta} = 30 \end{cases} \tag{23a}$$

$$\begin{cases} z_1^{2\alpha} = 20 \\ z_1^{2\beta} = -20 \end{cases} \tag{23b}$$

for each state $s = 1, 2$. Since the lost camel transaction is a zero-sum game, the payoff structure satisfies the win-lose condition. Then the lost camel transaction is not Shariah-compliant.

Here in order to derive the win-lose condition for non-zero-sum game, we consider a non-zero-sum game shown as Table 13.5 by

Table 13.5.   Modified Payoff of Lost Camel Transaction.

	The seller ($k = \alpha$)		The buyer ($k = \beta$)	
	Found	Not found	Found	Not found
Accept the contract	20	20	$-20 - 5 + 50 = 25$	$-20 - 5 = -25$
Reject the contract	0	0	0	0

modifying the lost camel transaction. The modified game assumes that the seller has no outside option in that the seller is not allowed to transact with any other buyers if the contract is not concluded with the specific buyer even if the camel is eventually found. If the contract is not concluded, both players gain zero regardless of the realized state of nature. In addition, the search cost borne by the seller $c = 5$ is considered. If the contract is concluded, players' payoff for each state of nature is calculated as

$$\begin{cases} z_1^{1\alpha} = 20 \\ z_1^{1\beta} = 25 \end{cases} \tag{24a}$$

$$\begin{cases} z_1^{2\alpha} = 20 \\ z_1^{2\beta} = -25. \end{cases} \tag{24b}$$

Hence the modified lost camel game is a non-zero-sum game. If the contract is concluded and $s = 1$ is realized, both players gain positive payoff, which is called a win-win relationship. However, the amount of payoff is not necessarily equal between the two players. If $s = 2$ is realized, the seller ($k = \alpha$) gains the positive payoff, whereas the buyer ($k = \beta$) gains the negative payoff, which exhibits the win-lose relationship. This example simply indicates the equality problem of the payoff structure under the non-zero-sum game. Under the non-zero-sum game, there are two possible cases: (1) the two players' payoffs show opposite signs or (2) the signs of two players' payoffs are the same, but their absolute values are not the same. Congruity in the signs of two players' payoffs implies the ex-post equality of payoff allocation for the non-zero-sum game. We call it "congruent sign condition" in payoff allocation. If the congruent sign condition in payoff allocation is satisfied, the payoff allocation is characterized by win-win relationship or lose-lose relationship. In addition to the congruent sign condition in payoff allocation, congruity in the absolute value of two players' payoffs implies the equity of payoff allocation. We call it "equitable allocation condition."

Ex-Post Equity Condition

Under the contract game Γ,

$$w^s = z_i^{s\alpha} + z_i^{s\beta} \, (s = 1, \ldots, S) \tag{25}$$

can be derived from the definition. If the contract game is characterized by the non-zero-sum structure, the equity condition of ex-post payoff allocation consists of (1) congruent sign condition and (2) equitable allocation condition. If the condition that the payoff z_1^{sk} for player α and player β is the same is represented as

$$z_1^{s\alpha} = \kappa^s w^s \tag{26a}$$

$$z_1^{s\beta} = (1 - \kappa^s) w^s \tag{26b}$$

$$0 < \kappa^s < 1 \tag{26c}$$

and if (26a)–(26c) are satisfied and state-contingent payoff sum w^s is positive, both $z_1^{s\alpha}$ and $z_1^{s\beta}$ are positive which implies a win-win relationship, whereas $w^s < 0$ implies a lose-lose relationship. Equitable allocation condition requires symmetric ratio of payoff allocation between the two players for any state of nature, i.e.

$$\kappa^1 = \cdots = \kappa^s = \kappa. \tag{27}$$

If κ^s is not symmetric for the state of nature s, relative payoff allocation depends on uncertain state of nature which implies "gambling." If (27) is satisfied, the ratio of payoff allocation κ does depend on uncertain state. Hence, it does not conflict with the concept of *maysir* too.

In the Appendix, the contract game of conventional *maysir* is analyzed. The structure of a conventional *maysir* transaction is a non-zero-sum game similar to the "lost camel." If the zero-sum structure of regret payoff is a necessary condition of *gharar* as proposed by Al-Suwailem (2000), *maysir* itself is *gharar*. However, Shariah prohibits *gharar* as well as *maysir* not only for the zero-sum payoff structure but also for containing a factor of gambling in a transaction. If players are rational and share common knowledge about the probability regarding the occurrence of state of nature, the ratio of payoff allocation must be consistent across the states shown as (27). Therefore, (27) can be interpreted as a condition which prohibits *maysir* in the broader sense as mentioned earlier. We call it "broad *maysir*."

Profit-Loss Sharing System

Shariah-compliant contract applies the profit-loss sharing system, where parties to a contract share the net profit or loss of a project. Under the complete contracting setting, the state contingent allocation of net payoff w^s can be completely described beforehand. Here we assume that the allocation is determined through bargaining between the rational players. Let w^s denote the payoff allocated to player α at state s. Under the contract type τ, the payoff of player k at the state s is represented by

$$z_1^{s\alpha}(\tau) = \tau^s \tag{28a}$$

$$z_1^{s\beta}(\tau) = w^s - \tau^s. \tag{28b}$$

If the contract is not concluded, the payoff of both players is zero. If both players have equal bargaining power, the Nash bargaining solution (Nash, 1950) gives the payoff of player α and β at the state s as

$$z_1^{s\alpha^*}(\tau) = \frac{w^s}{2} \tag{29a}$$

$$z_1^{s\beta^*}(\tau) = \frac{w^s}{2}. \tag{29b}$$

This discussion can be extended to the case where the bargaining power of players is not equal. The generalized Nash bargaining solution (Harsanyi, 1956; Harsanyi and Selten, 1972) between two players with the asymmetric bargaining power is written as

$$z_1^{s\alpha^*}(\tau) = \kappa w^s \tag{30a}$$

$$z_1^{s\beta^*}(\tau) = (1 - \kappa)w^s \tag{30b}$$

where κ denotes the parameter of bargaining power of player α. The parameter κ is the ratio of the net payoff allocated to player α.

The ratio of payoff allocation is determined endogenously through rational bargaining. Under such a setting, the ratio of payoff allocation κ does not depend on the state s, hence the contract where the net payoff is proportionally allocated is endogenously derived. Under this proportional allocation system, both players gain positive payoff if the net payoff $w^s > 0$, whereas negative if $w^s < 0$. Therefore, win-lose relationship cannot occur under the proportional allocation system. If a contract does not apply the proportional allocation system, win-lose relationship could occur at some state. Furthermore, since the ratio of payoff allocation κ is consistent across the states, it does not conflict with broad *maysir*. This proportional allocation system represented by (30a) and (30b) coincides with the profit-loss sharing system which is commonly used in Islamic finance.

Ex-Ante Efficiency Condition

Ex-post equity condition requires the relationship of players be either win-win or lose-lose at any state of nature ex-post. In addition, it requires the consistent ratio of net payoff allocation across the states, i.e. the proportional allocation system. If a transaction entails uncertainty, the net payoff of project is not always positive. If the net payoff is negative, parties to a contract shall share the loss. The information or knowledge of players are assumed to be symmetric. In this case, players are voluntarily motivated to join the contract only if the expected net payoff is positive. Hence, Pareto improvement of both parties' payoff guarantees voluntary conclusion of the contract, i.e.,

$$\sum_{s=1}^{S} p(s)w^s > 0. \tag{31}$$

The condition (31) is called "ex-ante efficiency condition" in this study. Under the proportional allocation system (30a) and (30b),

$$\sum_{s=1}^{S} p(s)z_1^{sk}(\tau) > 0 \ (k = \alpha, \beta) \tag{32}$$

is guaranteed if the contract satisfies ex-ante efficiency conditions. The above discussion can be summarized as **Proposition** below.

Proposition: under the complete contract setting, ex-post equity condition (26a)–(26c), (27) and ex-ante efficiency condition (31) are necessary conditions for Shariah compliance.

Proposition does not claim these two conditions are sufficient conditions for Shariah compliance. For example, any transaction associated with *riba* or *maysir* is not Shariah-compliant. The zero-sum game analyzed by Al-Suwailem (2000) cannot be Shariah-compliant according to ex-post equity condition and ex-ante efficiency condition.

A contract which does not satisfy ex-post equity condition creates a case of enmity where one party is rejoicing whereas the other feels regret. Prohibition of *gharar* intends to realize fair society by preventing transaction cases where somebody feels enmity.

Assessment of Shariah Compliance of Typical Contract

Conventional Islamic Contracts

Typical contracts associated with *gharar* are *munābadha, mulāmasa* and *hasat*. Table 13.6 summarizes the analytical result regarding Shariah compliance according to the two conditions we proposed above, i.e., ex-post equity condition and ex-ante

Table 13.6.   Assessment Result of *Gharar* for Conventional Transaction Cases.

	Shariah's position	Position A Knowledge asymetricity	Position A Existence of goods	Position B Enmity	Ex-post equity condition	Ex-ante efficiency condition	*Ghorm*	*Riba*
Lost camel	×	○	×	×	×	×	×	○
Munābadha	×	×	○	×	×	×	×	○
Mulāmasa	×	×	○	×	×	×	×	○
Hasat	×	○	×	×	×	×	×	○
Ju'alah	△	○	×	▲	▲	○	×	○
Salam	○	○	×	×	×	○	○	○
Murabaha	○	○	○	×	×	×	○	○

Note: ghorm indicates the risk of being unsold and of being defective during storage in commercial transactions (Rosly, 2005).

efficiency condition. The column of Shariah's position denotes the conventional positions of the Islamic jurisprudence concerning Shariah compliance in terms of *gharar*. "○" indicates that a contract is Shariah-compliant, whereas "×" indicates it is not. The columns of Position A and B indicate the conformity with the view of Position A and Position B discussed in the earlier section. Finally, the columns of ex-post equity condition and ex-ante efficiency condition indicate the conformity with these two conditions. According to Table 13.6, all contracts impermissible in the conventional Islamic jurisprudence do not satisfy the conditions that we propose.

Typical Non-Islamic Contracts

Two typical contracts in infrastructure projects, project finance contract and insurance contract are analyzed. These two contracts are commonly used in non-Islamic society, but they are not Shariah-compliant.

Project Finance Contract

A project applying the project financing scheme generally procures financial capital by stock and bond. Bond is viewed as being Shariah-incompliant since it is a financial instrument including interest rate, i.e., *riba* (Mohsin and Abbas, 1987). We apply the contracting game model to the project finance scheme to assess its Shariah compliance.

Let I be the amount of initial investment necessary for starting the project. vI is procured by issuing stocks and $(1 - v)I$ by issuing bonds. r represents the interest rate. Although the assumption of interest rate itself violates Shariah, here we focus on the rule of payoff allocation under the project finance scheme. The cash flow obtained from the project is stochastic. The cash flow is R if $s = s_H$ and 0 if $s = s_L$, where $R < rvI$. The payoff of stockholder ($k = \alpha$) and bondholder ($k = \beta$) at each state is written as

$$z_1^{s_H \alpha} = R - (1 - v)rI - vI \qquad (33a)$$

$$z_1^{s_H \beta} = (1 - v)rI - (1 - v)I \qquad (33b)$$

$$z_1^{s_L\alpha} = -vI \tag{34a}$$

$$z_1^{s_L\beta} = -(1-v)I. \tag{34b}$$

Obviously, payoff is not allocated proportionally across the states as the ratio of payoff allocation at state s_H is different from the one at state s_L. Therefore, the project financing scheme does not satisfy the ex-post equity condition and hence it is *gharar*. In general, bond is prohibited based on the prohibition of *riba*, i.e., interest rate. This analysis assumes that interest rate is seen as a contract scheme which defines the allocation of payoff. The rule of payoff allocation including interest rate does not allocate the payoff proportionally in any state.

Insurance Contract

Conventional insurance contract is not permissible since it entails *riba, maysir* and *gharar* (Al-Zuhayli, 2007). Conventional insurance is associated with *riba* because (1) the fund is invested in financial instrument with *riba* and (2) the allocation schedule of the revenue for the insurer and payment for the insuree.

The insured $(k = \beta)$ pays the fee τ to the insurer $(k = \alpha)$. A risk event which causes the damage D occurs with the probability p. The insurance is risk fair, i.e., $pD = \tau$. Both players are assumed to be risk-neutral differently from the conventional analysis of insurance where the insured is assumed to be risk-averse. If the risk event does not occur $(s = 2)$, the payoff of two players are written as

$$z_1^{2\alpha} = \tau \tag{35a}$$

$$z_1^{2\beta} = -\tau \tag{35b}$$

which obviously exhibits win-lose relationship. If the risk event occurs, the payoff of two players are written as

$$z_1^{1\alpha} = -D + \tau \tag{36a}$$

$$z_1^{1\beta} = -\tau \tag{36b}$$

which does not satisfy the proportional allocation of the net payoff and hence it is seen as *maysir*. Therefore the result of assessing Shariah compliance of conventional insurance based on the proposed model gives the same conclusion as the position of conventional Islamic jurisprudence.

Typical Shariah-Compliant Contracts

Two typical contracts deemed by Islamic jurisprudence as being Shariah-compliant, are *mushāraka* and *takaful*. *Musharākah* is a financing method based on the co-funding principle which can be an alternative for project finance projects in Islamic society. *Takaful* is an insurance system similar to the mutual insurance mechanism permitted in Islamic society. Shariah compliance of these two contracts are examined by applying the model we propose.

Musharākah

Musharākah is a transaction where multiple entities cooperate in a project by providing goods and labor forces as well as capital. Obtained surplus or loss is shared among the entities according to the predetermined ratio. Here we consider a case where a project finance project is carried out by applying *mushāraka*. Consider two players provide capital of which amount is I. The amount of capital provided by player α is μI and player β is $(1 - \mu)I$. Let $V^s (s = 1, 2)$ denote the amount of total cash flow obtained from the project. The amount of cash flow depends on the realized state of nature and $V^1 < V^2$ is assumed. The payoff of two players are written as

$$z_1^{s\alpha} = \mu(V^s - I) \tag{37a}$$

$$z_1^{s\beta} = (1 - \mu)(V^s - I) \tag{37b}$$

where (37) represents that the net payoff is allocated according to the profit-loss sharing scheme which satisfies the ex-post equity condition. In addition, if the project stakeholders share the same belief on the probability regarding the occurrence of state of nature, ex-ante efficiency condition is also satisfied.

Takaful

Takaful is a kind of insurance system where the insurance payment comes from a funding pool provided by the members. Consider a two possible state, the state where a risk event occurs ($s = 1$) and the state where no risk occurs until the expiration date ($s = 2$). If a risk event occurs, a member who suffered damage receives an amount of money that fully covers the damage. The money remaining in the funding pool is returned to the members.

Consider two players who face homogenous risk events. Two players, player α and player β provide the amount of money τ to a funding pool as donation. The funding pool is a funding source for compensating the damage of members. Players face a risk of which damage is D with the probability of occurrence p. We assume $2\tau = D$. In addition, we assume that the probability of the state where both players suffer a risk event at the same time is zero. All players know that each player shares the same knowledge and information. Since the money provided by the members is a donation, its ownership is deemed to have been waived. In fact, *takaful* is advocated as being Shariah-compliant based on the legal principle called *tabarru* which means donation (Gönülal, 2013). Hence, we define the timing after the members provide money as the initial state where the payoff is defined to be zero. If a risk event does not occur ($s = 2$), all money in the pool is returned to the members. Then the payoffs of player α and player β is written as

$$z_1^{2\alpha} = \frac{2\tau}{2} = \frac{D}{2} \tag{38a}$$

$$z_1^{2\beta} = \frac{2\tau}{2} = \frac{D}{2}. \tag{38b}$$

If a risk event occurs ($s = 1$), only player α suffers the damage D which is compensated by the pooled money. The remaining money

$2\tau - D$ in the pool is returned to each member. Then the payoffs of player α and player β is written as

$$z_1^{1\alpha} = \frac{2\tau - D}{2} = 0 \tag{39a}$$

$$z_1^{1\beta} = \frac{2\tau - D}{2} = 0 \tag{39b}$$

where (38) and (39) indicate the payoff allocation of *takaful* satisfies the ex-post equity condition. In addition, the expected payoff of two players are written as

$$E[z_1^{s\alpha}] = p\frac{D}{2} \tag{40a}$$

$$E[z_1^{s\beta}] = p\frac{D}{2} \tag{40b}$$

which obviously satisfies the ex-ante efficiency condition as well.

Discussion

Shariah-compliant contract requires satisfying the ex-post equity condition as well as the ex-ante efficiency condition. The ex-post equity condition concerns the manner of allocating the surplus brought by a project. On the other hand, the ex-ante efficiency condition concerns an issue which is also required in non-Islamic society. If we ignore the impact of the manner of payoff allocation on the incentives and long-term technological innovation, there is no difference between conventional contracting and Islamic contracting. Both contracting systems are efficient. In other words, the substantial difference between conventional contracting and Islamic contracting is associated with risk-sharing and the asymmetricity of knowledge.

Conventional contracting allows the asymmetricity of knowledge among the parties to a contract. The party with rich knowledge is supposed to bear the risk associated with that knowledge (Khan and Parra, 2003). On the contrary, the existence of asymmetric knowledge is regarded as *gharar* which is prohibited in Shariah. In Islamic society, sharing information and knowledge and cooperation among the stakeholders is required. The fruit of a project is distributed to the members in a fair manner. This kind of difference in contracting system is not a matter of relative merits, rather it is a matter of culture and religion.

Islamic contracting aims at contributing to the welfare of *ummah*. In order to realize the social justice in Islam, achieving economic growth and efficiency is not enough. Principles necessary for social justice in Islam are the principle of mutual responsibility and principle of social distribution (Muhammad Baqir, 1983). The principle of mutual responsibility prohibits pushing risk to a specific person and gaining wealth by doing so. On the other hand, the existence of knowledge and information is undeniable, which could contribute to the incentive of growth and innovation. The issue of incentive mechanism in Islam is beyond the scope of our study.

Conclusion

This chapter aims at theoretical analysis to derive the conditions to assess Shariah compliance of contracting. Shariah prohibits the concepts of *gharar* and *maysir*. We argue the intention of prohibition of *gharar* and *maysir* is to avoid the possibility of enmity regarding transaction by eliminating unnecessary uncertainty and risk. We propose two conditions, the ex-post equity condition and the ex-ante efficiency condition as necessary conditions for contracting being Shariah-compliant. Though there have been a number of studies to try defining those concepts rigorously, no consensus has been developed in practice so far. Note that the ex-post equity condition and the ex-ante efficiency condition can be the necessary conditions to exclude *gharar* in contracting, but they do not exclude all *gharar* contracting. Our contribution lies in accumulating a theoretical hypothesis for falsification to understand Shariah deeper.

Appendix

Maysir is referred in the following example of transaction to determine the allocation of meat blocks with different size by lottery. θ_1 denotes the choice to accept the contract and θ_2 denotes the choice to reject the contract. Player α owns two blocks of meat with different size. The market value of the larger block of meat is v_L and that of the smaller block of meat is v_S where $v_L > v_S$. Player β can purchase the right to obtain either of two blocks of meat by paying τ to player α. Assume $v_L > \tau > v_S$. The block of meat that player β obtains is determined by lottery. Let s_L denote the state where player β obtains the larger block and s_S the larger block. If this transaction does not realize, the payoff of players is written as

$$x_2^{s\alpha} = v_L + v_S$$

$$x_2^{s\beta} = 0$$

which means the payoff is independent from the result of lottery. On the other hand, if the transaction realizes, the payoff of players depending on the realized result of lottery is written as

$$\begin{cases} x_1^{s_L\alpha}(\tau) = \tau + v_S \\ x_1^{s_L\beta}(\tau) = -\tau + v_L \end{cases} \tag{1.1a}$$

$$\begin{cases} x_1^{s_S\alpha}(\tau) = \tau + v_L \\ x_1^{s_S\beta}(\tau) = -\tau + v_S. \end{cases} \tag{1.1b}$$

Hence, the net payoff of transaction for each player is calculated as

$$\begin{cases} z_1^{s_L\alpha}(\tau) = \tau - v_L \\ z_1^{s_L\beta}(\tau) = -\tau + v_L \end{cases} \tag{1.2a}$$

$$\begin{cases} z_1^{s_S\alpha}(\tau) = \tau - v_S \\ z_1^{s_S\beta}(\tau) = -\tau + v_S \end{cases} \tag{1.2b}$$

which implies that the transaction of *maysir* has zero-sum structure and win-lose condition holds. In addition, it can be proved that player α and β are motivated to enter the transaction only when their belief on the probability function regarding the lottery is inconsistent. Let $p_\alpha(s_L)$ and $p_\beta(s_L)$ denote the subjective belief of player α and β respectively regarding the probability that state s_L occurs. The condition that player α accepts the transaction is written as

$$p_\alpha(s_L)\{(\tau + v_S) - (v_L + v_S)\}$$
$$+ p_\alpha(s_S)\{(\tau + v_L) - (v_L + v_S)\} \tag{1.3}$$
$$= \tau - p_\alpha(s_L)v_L - p_\alpha(s_S)v_S > 0$$

and the condition for player β is

$$p_\beta(s_L)(-\tau + v_L) + p_\beta(s_S)(-\tau + v_S)$$
$$= -\tau + p_\beta(s_L)v_L + p_\beta(s_S)v_S > 0. \tag{1.4}$$

It is easily shown that if $p_\alpha(s_L) = p_\beta(s_L)$, (1.3) and (1.4) does not hold at the same time.

References

Allais, M (1953). Le comportement de l'homme rationnel devant de risque: Critique des postulates et axioms de l'ecole Americaine. *Econometrica*, 21, 503–546.

Al-Suwailem, S (2000). Towards an objective measure of *gharar* in exchange. *Islamic Economic Studies*, 7(1 & 2), 61–102.

Al-Suwailem, S (2006). *Hedging in Islamic Finance*. Islamic Research and Training Institute. Islamic Development Bank.

Al-Zuhayli (2007). Translated by M El-Gamal. *Financial Transactions in Islamic Jurisprudence*. Vol. 1 and 2. 2nd Ed. Dar Al-Fikl.

Badawi, Z (1998). The question of risk. *Islamic Banker*, 32, 16–17.

El-Gamal, M (2001). An economic explication of the prohibition of gharar in classical Islamic jurisprudence. *Islamic Economic Studies*, 8(2), 29–58.

El-Gamal, M (2006). *Islamic Finance: Law, Economics and Practice*. Princeton: Cambridge University Press.

Ellsberg, D (1961). Risk, ambiguity and the Savage axioms. *Quarterly Journal of Economics*, 75, 643–669.

Gönülal, SO (2013). *Takaful and Mutual Insurance*. The World Bank.

Harsanyi, JC (1956). Approaches to the bargaining problem before and after the theory of games. *Econometrica*, 24(2), 144–157.

Harsanyi, JC and R Selten (1972). A generalized Nash solution for two-person bargaining games with incomplete information. *Management Science*, 18(5), 80–106.

Imam Malik ibn Anas (2001). Translated by A Bewley. *Al Muwatta of Imam Malik ibn Anas: The First Formulation of Islamic Law*. Islamic Book Trust.

Kahneman, D and A Tversky (1979). Prospect theory: an analysis of decision under risk. *Econometrica*, 47, 263–291.

Khan, MFK and RJ Parra (2003). *Financing Large Project*. Pearson Prentice Hall.

Loomes, G and R Sugden (1982). An alternative theory of rational choice under uncertainty. *The Economic Journal*, 92(368), 805–824.

Mohsin, SK and M Abbas (1987). *Theoretical Studies in Islamic Banking and Finance.* Islamic Publications International.

Muhammad Baqir, al-Sadr (1961). Translated by MB Shubber. *Iqtisaduna; Our Economics.* Bookextra.

Nash, JC (1950). The bargaining problem. *Econometrica*, 18(2), 155–162.

Neumann, JV and O Morgenstern (1944). *Theory of Games and Economic Behavior.* Princeton University Press.

Rosly, S (2005). *Critical Issues on Islamic Banking and Financial Markets.* Kuala Lumpur: Dinamas Publishing.

Salanie, B (1997). *The Economics of Contracts: A Primer.* Cambridge, Massachusetts: The MIT Press.

Savage, LJ (1972). *The Foundations of Statistics.* 2nd Ed. New York: John Wiley and Sons.

Vogel, F and S Hayes, (1998). *Islamic Law and Finance: Religion, Risk, and Return.* Boston, Massachusetts: Kluwer Law International.

Chapter 14

Takaful for Construction Works' Contract: Concept and Application

PUTERI NUR FARAH NAADIA Mohd Fauzi
and KHAIRUDDIN Abdul Rashid

Introduction

In *muamalat*, risks refer to potential loss in a transaction (Asyraf Wajdi Dusuki, 2012). Literally, in the context of Islamic financial transactions, risks imply several meanings such as exposure of loss or fear of destruction due to unpredictable events in future (Asyraf Wajdi Dusuki, 2012). In Islam, Muslims are required to anticipate the potential of loss that might happen in the future and take utmost precautions to mitigate them. This is to encourage one to improve his life and achieve his best in this world and the hereafter.

The concepts of anticipation and mitigation of risks are enshrined in the Quran:

> "Then, there shall follow seven hard years (of drought) which will consume all but little of that which you have stored (as seeds)." (Quran 12: 48).

In addition, the Quran also emphasizes on the significance of anticipation and mitigation of risks:[1]

> "Then he said, "O my sons, do not enter (the city of Egypt) through one gate, but go into it by different gates. However I cannot (with this advice) be of any help to you against whatever decree decided by Allah; judgment rests with Allah alone. In Him, I have put my trust. And in Him alone let the trustful put their trust." (Quran 12: 67).

Moreover, the concept of anticipation and mitigation of risks is based on a *hadith* where Prophet Muhammad (pbuh) was asked by a Bedouin whether he should tie his camel and leave it to Allah (*s.w.t*) or completely leave it to the will of Allah (*s.w.t*). The Prophet Muhammad (pbuh) instructed him to leave the fate of his camel to Allah (*s.w.t*) only after he ties it:

> "Anas bin Malik narrated that a man said: 'O Messenger of Allah! Shall I tie it and rely (upon Allah), or leave it loose and rely (upon Allah)?' He said: 'Tie it and rely (upon Allah).' (Tirmidhi, Vol.4, Chap. 60, No. 2517).

The *hadith* demonstrates that striving and making best efforts before complete submission to Allah (*s.w.t*) is important.

Specifically for construction, the extensive and complex tasks of physically assembling a building expose construction works to risks. Any occurrence of risks might involve a large amount of loss to the contracting parties. Therefore, mitigation of risks in construction is essential to ensure works are protected. Conventionally, there are several methods to mitigate risks as described in Table 14.1.

In Islam risks are to be managed through risk-sharing (Department of Islamic Development Malaysia [JAKIM], 2009, 1972). Risk-sharing refers to individuals in a group mutually agreeing to assist and protect each other (Ahmad Mazlan *et al.*, 2012). This method is applied through a protection scheme i.e. *takaful* or Shariah compliance insurance.

This chapter discusses the concept and application of *takaful* in the context of the construction industry. Following the introduction, this chapter presents the concept and principles of *takaful*, its origin

---

[1] Advises of Prophet Ya'akob (pbuh) to his children when they entered Egypt.

Table 14.1.   Conventional Risk Mitigation Methods.

Method	Description
Risk avoidance	Conscious effort by an individual or firm to avoid exposure to a particular risk. However, by doing so, an individual or firm loses out on the potential profit when no risk occurs.
Risk reduction	Risk reduction also refers to loss control reduction that includes techniques to reduce the potential severity of losses that do occur. For instance, a two-hour fire rated door is designed and constructed to allow time to put out a fire that may reduce the risk of loss by fire.
Risk retention	The process of retaining the risks where the risks are already at an acceptable level.
Risk transfer	Transferring or shifting risks, which are not avoided or retained through a contract or an agreement. There are two ways of transferring a risk, (1) through an insurance contract in which a contracted compensation is paid to the insured by the insurer in the event of risk occurrence and (2) through a non-insurance transfer that refers to methods other than insurance by which a risk is transferred to another party.
Risk-sharing	The risks are shared with others in the same group. In the event of loss and damage occurring from insurable risks, the loss and damage will be shared accordingly.

Source: Ahmad Mazlan *et al.* (2012); Engku Rabiah Adawiah and Odierno (2010).

and development in general and specifically in Malaysia. This is followed by discussions on the application of *takaful* in construction, using the Malaysian PWD 203A standard form of contract as basis for the discussion. Finally, the chapter ends with concluding remarks.

Concept and Principles of *Takaful*

Literally, the term *"takaful"* is derived from an Arabic root word *"kafl"* which means guarantee, responsibility or an act of securing one's need. The Islamic Financial Services Act 2013 (IFSA 2013) interprets *takaful* as: "An arrangement based on mutual assistance under which *takaful* participants agree to contribute to a common fund providing for mutual financial benefits payable to the *takaful* participants or their beneficiaries on the occurrence of pre-agreed events."

The principles of *ta'awun* and *tabarru'* form the operation of *takaful*. *Ta'awun* and *tabarru'* refer to the principle of mutual assistance and contribution in a form of donation respectively. Islam encourages the principle of mutual assistance among people. This principle is enshrined in the Quran:

> "...Help one another in furthering virtue and God-consciousness, and not in what is wicked and sinful. And remain conscious of Allah, for He is stern in retribution (to those who disobeyed His orders)." (Quran 5: 2).

In addition, Prophet Muhammad (pbuh) preached that people help each other to relieve their burdens as narrated by 'Abdullah bin Umar:

> "A Muslim is a brother of another Muslim, so he should not oppress him, nor should he hand him over to an oppressor. Whoever fulfilled the needs of his brother, Allah will fulfill his needs; whoever brought his (Muslim) brother out of a discomfort, Allah will bring him out of the discomforts of the Day of Resurrection, and whoever screened a Muslim, Allah will screen him on the Day of Resurrection." (Sahih Bukhari, Vol. 3, No. 622).

Takaful donation (*tabarru'*) is a contribution made by participants into a *takaful* fund (IFSA, 2013). The fund is managed by an operator and to be utilized in assisting the participants in the event of occurrence of any specified and agreed event (IFSA, 2013). If there is no payment to claim made by participants, the amount of money in the fund will be distributed accordingly based on an agreed portion. The principle of *tabarru'* is not aiming to profit from uncertainties but to fulfill obligations of mutual protection and assistance (Mohd Ma'sum, 2007).

In addition, the fund is allowed to be invested in any Shariah-approved investment to generate income (Rodziah and Zairol Azhar, 2012). Profit from the investment shall be credited back to strengthen the *takaful* fund (Zubair, 2014).

Practice of Takaful and Conventional Insurance

In *takaful*, an operator is required to have a Shariah Advisory Board. The Shariah Advisory Board comprises of Shariah experts and scholars to ensure that the activities conducted by the operator in managing the fund are in compliance with the Shariah (Takaful Act 1984).

In the practice of *takaful*, identified risks are shared among the *takaful* participants (International Shariah Research Academy for Islamic Finance (ISRA), 2012). *Takaful* operator which has been appointed to manage a *takaful* fund will ensure that *takaful* participants pay an equal amount of *tabarru'* and get equitable compensation in the occurrence of any loss and damage (Engku Rabiah Adawiah and Odierno, 2010).

In contrast, under a conventional insurance, an insured enters into a sell-purchase agreement in order to buy an insurance policy where the insured agrees to pay a fixed premium and in return, the insurer agrees to bear risks and pay compensation in the event of loss or damages which are covered under the policy (Insurance Council of Australia, 2013). Through the payment of an insurance premium, the risk borne by the insured is transferred to an insurance company.

In terms of contractual relationship, a conventional insurance contract is based on an exchange type of contract that offers a sell and purchase agreement between an insurer and insured (Engku Rabiah Adawiah and Odierno, 2010). The insurance company sells a guarantee in the form of insurance policy to a prospective insured for a premium. Upon signing of an insurance policy, the insured will be provided with an insurance policy that entitles him to claim in the event of loss and damage arising from the specified and agreed events. Conversely, there is no buying and selling of policy practiced in a *takaful* contract (Engku Rabiah Adawiah and Odierno, 2010; Mohd Ma'sum, 2007).

Table 14.2 provides comparison of *takaful* and conventional insurance.

Origin of *Takaful*

Figure 14.1 shows the key events on the origin and evolution of *takaful*.

Table 14.2.   Comparison of *Takaful* and Conventional Insurance.

Concept and principles	Forms of insurance	
	Takaful	Conventional insurance
Risk	Sharing risks between the same group	Transfer risks to one party
Modus operandi	Based on mutual cooperation (*ta'awun*) and donation (*tabarru'*)	Based solely on commercial factors
Business operations/ activities	Free from the Shariah prohibitory elements i.e. uncertainty (*gharar*), gambling (*maysir*) and usury (*riba*)	Presence of the Shariah prohibitory elements i.e. uncertainty (*gharar*), gambling (*maysir*) and usury (*riba*)
Payment of subscription	Contributions payment in the form of donations through donation contract (*tabarru'*)	Premium payment which creates an obligation against the insurer on a sale and purchase contract

Source: Engku Rabiah Adawiah and Odierno (2010).

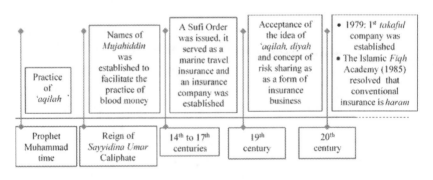

Figure 14.1.   Key Events on the Origin and Evolution of *Takaful*.
Source: Mohd Ma'sum (2007); International Shari'ah Research Academy for Islamic Finance (ISRA) (2012).

With reference to Figure 14.1, the origin of *takaful* is based on the practice of *'aqilah*,[2] which is an arrangement of mutual help at the

---

[2]A paternal relatives killer — A practice where the ancient Arab tribes had to be prepared to compensate on behalf of a killer to the heir of the victim (Al-Bukhari [2009]. *Sahih Bukhari, Kitab Diyat* [Book of Blood Money], Vol. 3, No. 45, p. 1543).

time of Prophet Muhammad (pbuh). *'Aqilah* refers to the practice whereby people of a given tribe would seek financial assistance from one of its members if he faces an unexpected liability such as paying for the blood money (*diyah*)[3] (Mohd Ma'sum, 2007).

Subsequently, during the period of the second caliph, Saidina Umar (R.A), he encouraged the practice of *'aqilah*. The names of *Mujahiddin*[4] from all districts of the States were recorded to facilitate the contribution of *diyah* in the event of manslaughter committed by anyone of their own tribe (ISRA, 2012).

During the period between 14th to 17th centuries, a Sufi Order was issued and served as a form of marine travel insurance and an insurance company was established by *Kazeruniyya*.[5] Merchants made payments that were predetermined by the Order as consideration to protect them against perils of sea voyages.

In the 19th century, Ibn 'Abidin, a Hanafi lawyer proposed an opinion that led other Muslims to accept the idea of *'aqilah, diyah* and the concept of risk-sharing as not only a customary practice but also as an insurance business (Mohd Ma'sum, 2007).

The first *takaful* company, Islamic Insurance Company was established in 1979 by the Dar al Maal Al Islami group in Sudan and in 1983, the Islamic *Takaful* Re *Takaful*, Bahamas was established (Rodziah and Zairol Azhar, 2012). In addition, *fatwas* on the legality of Islamic insurance practices were established. The Islamic *Fiqh* Academy (1985) emanating from the Organization of Islamic Conference (OIC), meeting at its Second Session in Jeddah on December 22–28, 1985 resolved that conventional insurance is *haram* according to the Shariah.

---

[3]Compensation to be paid by *'aqilah* to the victim's family (Uswatun Hasanah, 2011).

[4]*Saidina Umar* (R.A) commanded that a list of Muslim brothers-in-arms be established and those names had to be responsible to contribute blood money (Mohd Ma'sum Billah, 2007).

[5]An order that was established by a Sufiism school of *Kazeruniyya* Abu Ishaq Ibrahim Ibn Shahriyah (963–1035 CE). The Order was active in the port cities along the coast of India and China (Mohd Ma'sum Billah, 2007).

History of Takaful in Malaysia

In Malaysia, the awareness of *takaful* began in June 1972 when the National Fatwa Committee of the Malaysian Islamic Affairs Council issued a decree that the conventional insurance is prohibited or *haram*. This is due to the presence of elements of uncertainty (*gharar*), the involvement of interest (*riba*) and the presence of gambling (*maysir*) in conventional insurance transactions.[6]

> "As for those who take usury, they shall not be able upright but shall rise up like one whom Satan has demented by his touch, for they claimed that: "Trade is like usury." Whereas Allah has permitted trading and forbidden usury..." (Quran 2: 275).

> "Do not eat up your property among yourselves by unjust means, nor use it as bait for the judges in order that you may knowingly (and wrongfully) commit sin by eating up a part of other people's property." (Quran 2: 188).

> "O believers! Wine and gambling, idols, and divining arrows are (all of them) abominations devised by Satan. Avoid them, so that you may prosper." (Quran 5: 90).

A consistent decree has also been issued by the National Fatwa Committee of the Malaysian Islamic Affairs Council (2009) that the current implementation of *takaful* in Malaysia is in line with the Shariah. In addition, the National Fatwa Committee of the Malaysian Islamic Affairs Council in its meeting on December 14–16, 2009 issued a decree that Muslims should choose an insurance system based on the Shariah i.e. *takaful*.

In 1982, a special committee called *"Badan Petugas Khas"* was set up by the government to study the feasibility of setting up Islamic insurance in Malaysia. The committee concluded that the concept of insurance is accepted but opposed the involvement of prohibition elements in conventional insurance practices.

Subsequently, in 1985, the Takaful Act 1984 — a legal basis and source of *takaful* legislation in Malaysia — was enforced to regulate *takaful* business (Central Bank of Malaysia, 2005). The Act

---

[6]Department of Islamic Development Malaysia (JAKIM) (1972). Insurance.www.e-fatwa.gov.my [Accessed October 21, 2014].

was formed to ensure *takaful* operations embodied principles and requirements of the Shariah. The Act contains rules and regulations governing *takaful* businesses (Central Bank of Malaysia, 2013, 2005).

The first *takaful* company, Syarikat Takaful Malaysia (STM) Bhd was formed in 1984. The Malaysian *takaful* industry continued to grow with the establishment of other *takaful* operators from 1993 to 2000. In addition, the ASEAN Retakaful International (L) Ltd. was established in 1997 to facilitate *retakaful* arrangements among *takaful* operators in Malaysia and in the region, i.e. Brunei, Singapore as well as Indonesia (Ernst & Young Global Malaysia, 2016).

In 2001 the Financial Sector Masterplan (FSMP) was introduced, with the objective to enhance the capacity of the *takaful* operators and strengthen the legal, Shariah and regulatory frameworks (Central Bank of Malaysia, 2015). In 2002, the Malaysian Takaful Association was established. Its objectives include to promote the development of the *takaful* industry in Malaysia.

Towards realizing the aspiration of Malaysia becoming an international center for Islamic finance, the Central Bank introduced the Financial Sector Blueprint 2011–2020. In addition, the Takaful Operating Framework and revision of the Shariah Governance Framework were issued in 2012. In 2013, the enactment of Islamic Financial Services Act (IFSA) was announced to provide a stronger legal foundation to spur the growth of Islamic finance sector including *takaful* business.

In 2014, the bank issued completed revisions on enforcement of risk-based capital for *takaful* that support a strong and independent role of appointed actuary in promoting a sound management of *takaful* and financial risks (Central Bank of Malaysia, 2016). The Internal Capital Adequacy Assessment Process (ICAAP) for *takaful* operators was introduced in 2015 with aims to promote a rigorous process for capital management which is aligned with the risk profile of each *takaful* operator. Figure 14.2 shows the key development of the Malaysian *takaful* industry focusing on the year 2010 till present.

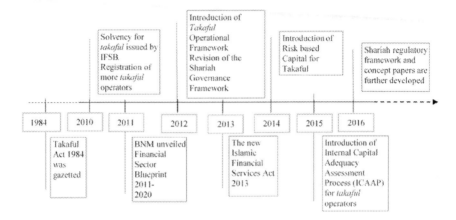

Figure 14.2.   Key Development of *Takaful* in the Malaysian *Takaful* Industry.
Source: Central Bank of Malaysia (2015, 2016).

Takaful Models

There are four most commonly employed *takaful* models i.e. profit-sharing (*mudharabah*) model, agency (*wakalah*) model, hybrid model and endowment (*wakaf*) model (Azman, 2013; Sheila Nu Nu and Syed Ahmed, 2013; Zubair, 2014).

1. Profit-sharing (*mudharabah*) model
    It operates as a profit-sharing arrangement between *takaful* participants and operator. In this model, a *takaful* operator acts as an entrepreneur or manager (*mudharib*) that manages the operation of the *takaful* fund while *takaful* participants act as capital providers (*rabul-mal*) that provide *tabarru'* to the *takaful* fund. The *takaful* fund is invested in Shariah-compliant investments and profit therefrom will be distributed between *takaful* participants and operator on a pre-agreed ratio upon inception of the contract. In addition, the deficit of the *takaful* fund shall be borne solely by the *takaful* participants on the condition that *takaful* operator is absolved from any misconduct or negligence.
2. Agency (*wakalah*) model
    This refers to an agreement between *takaful* participants as the master or principal (*muwakkeel*) and *takaful* operator as an agency

or a representative (*wakeel*) to manage the fund for a defined fee. Any liabilities of risks underwritten and surplus arising from the fund are borne by and belong exclusively to the *takaful* participants.

3. Hybrid model

   This is a combination of agency and profit-sharing (*wakalah-mudharabah*) model. In a hybrid model a *wakalah* model is used for underwriting whilst a *mudharabah* model is used for investment purpose.

4. Endowment (*wakaf*) model

   Permanent dedication by Muslims of any property for religious, pious or charitable purposes as recognized under the Shariah. In relation to this, the *wakaf* is deemed as *tabarru'*.

   The application of hybrid *takaful* model is adopted officially in Bahrain and also recommended by the Accounting and Auditing Organization for Islamic Financial Institutions (AAOIFI) (Zubair, 2014). Most operators in Pakistan, Saudi Arabia, UAE, South Africa, Qatar and Malaysia also apply the hybrid model for *takaful* business (Syarikat Takaful Malaysia Sdn. Bhd., 2014).

Main Types of Takaful

Notwithstanding the models being used, there are two main types of *takaful* i.e. family and general:

1. Family *takaful*

   It provides protection to participants' life and health through a combination of long-term investment and mutual financial assistance scheme. In addition, family *takaful* may also include protection for education, mortgage, health and riders, travel and endowment (*wakaf*), annuity as well as investment-linked benefits.

2. General *takaful*

   It provides protection to participants against loss and damage of properties or assets i.e. motor, fire, marine, engineering and construction, and aviation loss or damage resulting from a catastrophe or disaster inflicted upon real estate, assets or belongings of participants.

Application of *Takaful* for Construction Works

In the practice of construction, it is a condition precedent to the commencement of the works for a contractor to insure the works against loss and damage as prescribed by the major standard forms of construction contracts. For example, in Malaysia the protection of construction works and the requirement to secure an insurance are stated in clause 18.0 (Insurance of works) and 48.0 (Defects after completion) of the PWD 203 & 203A Standard Form of Contract (Rev. 1/2010).[7] In addition, there are two other types of insurance required under the standard form of contract namely, (1) insurance against personal injuries and damage to property[8] and (2) Workmen Compensation.[9]

Specifically for protection of works, the insurance must be taken up to the full value of the works. Figure 14.3 identifies the key events relevant to the insurance of works under the PWD 203.

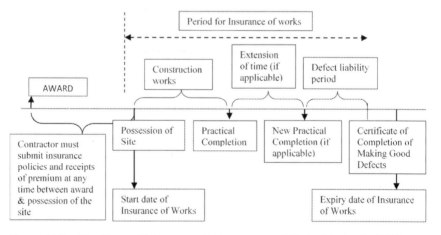

Figure 14.3.   Key Events Relevant to the Insurance of Works Under the PWD 203.

---

[7]PWD is a series of a suit of standard form of contract predominantly used for construction and public works contract in Malaysia.
[8]Clause 14.0 and 15.0 of PWD 203 & 203A (Rev. 1/2010).
[9]Clause 16.0 and 17.0 of PWD 203 & 203A (Rev. 1/2010).

A contractor has an option to insure works using either the conventional insurance or *takaful*. A combined policy comprising of insurance of works and third party is also available and this is known as Contractor's All Risks (CAR).

In relation to insurance of works, based on the comparative analysis of the current CAR, the practice does not differentiate between *takaful* and conventional insurance. However, both forms of insurance differ in terms of concepts and principles. Table 14.3 provides a summary of comparative analysis in terms of the main provisions between a typical CAR *takaful* scheme and CAR insurance policy.

Table 14.3. Summary of Comparative Analysis in Terms of the Main Provisions Between a Typical CAR *Takaful* Scheme and CAR Insurance Policy.

Main provisions		CAR *takaful*	CAR insurance
Scope of cover	Section 1: Protection to works	√	√
	Section 2: Protection to third party	√	√
Covered parties	Government or Employer as the principal	√	√
	Main contractor	√	√
Sum insured	Equal to total construction sum	√	√
	Additional of the contract sum	√	√
Period of cover	Site possession to Practical Completion	√	√
	Practical Completion to Certificate of Completion of Making Good Defects (CCMGD)	√	√
Loss and damage	As specified in construction contract	√	√
	Exclude risks under exclusion and special clauses	√	√
General exclusions	War, invasion, act of foreign enemy, hostilities, rebellion, revolution, insurrection, mutiny, lock-out, military, conspiracy, confiscation, commandeering, requisition	√	√
	Nuclear reaction, radiation or radioactive contamination	√	√

(Continued)

Table 14.3.　(*Continued*)

Main provisions		CAR *takaful*	CAR insurance
	Willful act or willful negligence of the Insured or of his representatives	√	√
	Cessation of work	√	√
Special conditions	Underground cables, pipes and other facilities	√	√
	Fire-fighting facilities and fire safety on construction site	√	√
	Safety measures with respect to precipitation, flood and inundation	√	√
Additional contribution/ premium	Riot, strike, civil commotion	√	√
	Airfreight	√	√
	Vibration, removal or weakening of support	√	√
	Earthquake, flood and inundation (subject to terms, exclusion, provisions, conditions and endorsement)	√	√
	Crops, forests and cultures (subject to terms, exclusion, provisions, conditions and endorsement)	√	√
Operation	Compliance with the requirements of the Shariah (non-involvement of uncertainty (*gharar*), gambling (*maysir*) and usury (*riba*)	√	X
	Surplus share condition	√	X
	Wakalah fee	√	X

Source: Etiqa Insurance and Etiqa Takaful Sdn. Bhd. (2014).

Conclusion

The application of *muamalat* in business transactions, including in the construction industry, is now a common phenomenon and is becoming a rising trend (Central Bank of Malaysia, 2016).

Fundamentally, the presence of the elements of *gharar* (uncertainty), *maysir* (gambling) and *usury* (interest or profiteering) is the reason why conventional insurance is forbidden or *haram*, hence making it non-Shariah-compliant. Through *takaful*, the application of *ta'awun* and *tabarru'* (mutual and voluntary donation) eliminates the presence of *gharar*, *maysir* and *usury*, hence making *takaful* Shariah-compliant.

In Malaysia, both insurance schemes i.e. conventional and *takaful* are available to the general population. The conduct of their respective businesses is robustly governed by the relevant agencies. Consequently, *takaful* offers a credible alternative to conventional insurance.

However, to the Muslims *takaful* offers an advantage in terms of providing them with the opportunity to conduct their worldly affairs that meets the requirements of the Shariah. This aspect is important as conducting their daily affairs including business transactions that meet the requirements of the Shariah is treated as an act of worship or *ibadah* to the Muslims.

The late debut of *takaful* has resulted in its application in the construction industry trailing behind conventional insurance and hence, the scarcity of literature on the subject of *takaful* in construction. This chapter hopes to inspire more works and publication to be done in the area of *takaful* for the construction industry.

List of Legislation

Islamic Financial Service Act (IFSA) 2013 (Act 759)
Takaful Act 1984 (Act 312)

References

Ahmad Mazlan, Z, Badrul Hisham Abd. Rahman, Nasser Yassin and Jamil Ramly (2012). *Basic Takaful Practices: Entry Level for Practitioners.* Kuala Lumpur, Malaysia: Islamic Banking & Finance Institute Malaysia (IBFIM) & Malaysian Takaful Association.

Al-Bukhari, Abu Abdullah Muhammad bin Ismail bin Ibrahim bin al-Mughira al-Ja'fai (2009). Sahih bukhari (d. 256/870). In *Sahih al-Bukhari.* Mika'il al-Almany (ed.) Translated by M Muhsin Khan.www.islamspirit.com

Asyraf Wajdi Dusuki (2012). Principles and application of risk management and hedging instruments in Islamic finance. www.iefpedia.com/english/wp-content/uploads/2012/07/Asyraf.pdf [Accessed March 24, 2017].

Azman, MN (2013). A Shariah compliance assessment on Takaful investment link. http://drazman.net/wp-content/uploads/2013/03/Dr.Azman-Takaful-inv estment-link-assesment-final-0305091.pdf [Accessed December 10, 2014].

Central Bank of Malaysia (2005). *Concept and Operation of General Takaful in Malaysia (Special Article)*. Kuala Lumpur: Central Bank of Malaysia.

Central Bank of Malaysia (2013). *Guidelines of Takaful Operational Framework*. BNM Guidelines & Circulars: www.bnm.gov.my [Accessed January 10, 2014].

Central Bank of Malaysia (2015). *Takaful*: Growing from strength to strength. Malaysia: World's Islamic Finance Market Place.

Central Bank of Malaysia (2016). *Takaful* annual report. www.bnm.gov.my [Accessed December 9, 2014].

Department of Islamic Development Malaysia (JAKIM) (1972). Insurance. http://e-muamalat.islam.gov.my [Accessed November 2, 2018 in Bahasa Melayu/Malay].

Department of Islamic Development Malaysia (JAKIM) (2009). Conventional insurance law to promote products by Shariah perspective. http://e-muamala t.islam.gov.my/ms/ [Accessed November 2, 2018: in Bahasa Melayu/Malay].

Engku Rabiah Adawiah, EA and Odierno (2010). *Panduan Asas Takaful: Panduan tentang falsafah dan prinsip Takaful berdasarkan Shari'ah*. Kuala Lumpur, Malaysia: CERT Publications Sdn. Bhd.

Ernst & Young Global Malaysia (2016). *Malaysian Takaful Dynamics*. Central Compendium 2015. Malaysia: Ernst & Young Global Malaysia.

Government of Malaysia (2007). Standard Form of Contract to Be Used Where Bills of Quantities Form Part of the Contract. Public Works Department (PWD) Form 203.

Government of Malaysia (2010). Standard Form of Contract to Be Used Where Bills of Quantities Form Part of the Contract. Public Works Department (PWD) Form 203A (Rev. 1/2010).

Etiqa Insurance and Etiqa Takaful Sdn. Bhd. (2014). *Takaful* products. www.etiqa. com.my [Accessed December 17, 2014].

Insurance Council of Australia (2013). http://www.insurancecouncil.com.au [Accessed November 19, 2013].

International Shari'ah Research Academy for Islamic Finance (ISRA) (2012). *Islamic Financial System: Principles & Operations*. Kuala Lumpur, Malaysia: Pearson Custom Publishing.

Islamic Fiqh Academy (1985). *Resolutions and Recommendations of the Council of the Islamic Fiqh Academy 1985–2000*. Jeddah, Saudi Arabia: Organization of Islamic Conference of Islamic *Fiqh* Academy.

Mohd Ma'sum, B (2007). *Applied Takaful and Modern Insurance. Law and Practice*. Selangor, Malaysia: Sweet & Maxwell Asia.

Rodziah, A and A Zairol Azhar (2012). *Takaful*. Kuala Lumpur, Malaysia: Pearson Malaysia Sdn Bhd.

Sheila Nu Nu, H and S Syed Ahmed (2013). Shar'ah and ethical issues in the practice of the Modified Mudharabah family *takaful* model in Malaysia. *International Journal of Trade, Economics and Finance*, 4(3), 340–342.

Syarikat Takaful Malaysia Sdn. Bhd. (2014). Glossary of *Takaful*. www.takaful malaysia.com.my [Accessed October 21, 2014].

Tirmidhi, Muhammad lbn Eisa At-Tirmidhi (2007). English translation of Jami' at-Tirmidhi (d. 279/892). In *Jami' at-Tirmidhi*. Hafiz Abu Tahir Zubair 'Au Za'I (ed.) Translated by Abu Khaliyl, p. 509. Riyadh: Makktaba Dar-us-Salam.

Uswatun Hasanah (2011). Insurance and Islamic law. *Indonesian Journal of International Law*, 9(1), 133–151.

Zubair, H (2014). *Islamic Banking and Finance: An Integrative Approach*. Shah Alam, Selangor, Malaysia: Oxford Fajar Sdn. Bhd.

Chapter 15

Hibah Mu'Allaqah (Conditional Gift) and Its Application in *Takaful*

AZMAN Mohd Noor

Introduction

Takaful is an alternative Shariah-compliant protection for the construction industry. The related products can be Contractor All Risks (CAR) *Takaful*, Group *Takaful*, Key Person Coverage *Takaful*, Fire *Takaful* and other related family *Takaful* which are Shariah-compliant alternatives to the conventional insurance.

This chapter proposes the concept of conditional *hibah* as an alternative instrument to resolve some issues in *takaful*, i.e the basis for paying *takaful* benefit absolutely to the nominated nominee as suggested by the new Islamic Financial Service Act in Malaysia 2013.[1] There has been a dispute among some contemporary scholars whether this practice is in line with Shariah since the *takaful* participants do not own the benefit during their lifetime as it does not exist yet.[2]

---

[1]Can be accessed at http://www.bnm.gov.my/documents/act/en_ifsa.pdf.

[2]Earlier related research was presented at ISRA Islamic Finance Seminar 2008 with the title "Ownership and Hibah Isssues of the Takaful Benefit. The paper can be accessed at http://www.iefpedia.com/english/wp-content/uploads/2009/10/Ownership-and-Hibah-Issues-in-Takaful-Classical-Discourse-and-Current-Implementation.pdf.

Definition of *Hibah Mu'allaqah* and the Juristic Opinions

According to juristic scholars, *hibah* is defined as *"tamlik al-'ain bi dun 'iwad"* (Al-Sharbini, 2000), which means transferring ownership without any consideration. *Hibah mu'allaqah* or *ta'lik al-hibah* is a conditional gift that is subject to the happening of a particular event. Generally, with reference to this type of *hibah*, its contract is only valid and binding once the conditions are fulfilled. However, there are different opinions between Muslim jurists on the validity of conditional *hibah*, which can be divided into two major opinions.

First Opinion: The Advocates of Hibah Mu'allaqah

This is the opinion of Maliki, Hanafi and some of the Hanbali schools of thought, specifically the opinion of Ibn Taimiyah and his disciple Ibn al-Qayyim. They are of the opinion that conditional *hibah* is valid and binding upon the fulfilment of its condition. This section will discuss the opinions or views from these schools of thought which allow conditional *hibah*.

Firstly, the opinion of Hanbali school of thought which allows *hibah* with conditions is clearly stated in the book *al-Insaf'* (Al-Mardawi, 1997)[3] which narrates the opinion of Imam Ahmad:

ظاهر كلام الإمام أحمد في رواية أبي الحارث صحة دفع كل واحد من الزوجين إلى الآخر مالاً على أن لا يتزوج .. ومن لم يف بالشرط لم يستحق العوض، لأنها هبة مشروطة بشرط فتنتفي بانتفائه انتهى .

"It is considered valid if a husband and wife paid (gave *hibah*) to one another with a condition that the husband should not marry a second wife but, if the condition is not fulfilled, the *hibah* cannot be claimed, (hence, the husband is not entitled to get the *hibah*). This is because the offer for *hibah* is conditional which lapses in the absence of the fulfilment of the condition."

Ibn Taimiyyah preferred the opinion which allows a conditional *hibah* with any kind of condition (*ta'lik hibah*) (Al-Mardawi, 1997).[4] His student, Ibn Qayyim refuted the opinion that does not allow

---

[3] Al-Mardawi, Ali bin Sulaiman, *al-Inshof fi Makrifah al-Rajih min al-Khilaf ala Mazhab al-Imam Ahmad bin Hanbal*, (Bayrut: Manshurat Muhammad 'Ali Baydun, 1997), vol. 20, p. 391.
[4] Al-Mardawi and Ali bin Sulaiman (1997). vol. 17, p. 44.

conditional *hibah*. He argued that the prohibition of this contract was not supported by any evidence from the Quran, *hadith* or even *ijma'* of the Prophet's companions. He challenged anyone to bring any evidence which clearly states that *hibah* will become void because of pending conditions. On the other hand, in supporting his argument, Ibn Qayyim asserted that, there is clear evidence that proves the Prophet (pbuh) made a conditional *hibah*, by referring to the following *hadith* which was narrated by Jabir:

قال : لو قد جاء مال البحرين لأعطيتك هكذا وهكذا ثم هكذا ثلاث حثيات . وأنجز ذلك له الصديق رضي الله عنه لما جاء مال البحرين بعد وفاة رسول الله صلى الله عليه وسلم.

(The Prophet) said to Jabir: "If the wealth (disembarks) from Bahrain, I will give you such and such amount (of money), with three grasp." Abu Bakar al-Siddiq fulfilled the promise of the Prophet (pbuh) (after the demise of the Prophet) to give *hibah* after the occurrence of the condition, when the wealth from Bahrain arrived.[5]

By referring to the above evidence, Ibn Qayyim suggests that *hibah mu'allaqah* is similar to a promise. In his book *Ighathat al-Lahfan*, he describes the following:

فإن قيل : كان ذلك وعداً . قلنا : نعم ، والهبة المعلقة بالشرط وعد ، وكذلك فعل النبي صلى الله عليه وسلم لما بعث إلى النجاشي بهدية من مسك وقال لأم سلمة : إني قد أهديت إلى النجاشي حلة وأواقي من مسك ولا أرى النجاشي إلا قد مات ولا أرى هديتي إلا مردودة فإن ردت علي فهي لك . وذكر الحديث . رواه أحمد .

"If someone said that (a *hibah*) amounts to a promise (*wa 'ad*), we said, yes, a *hibah* that is conditional with a condition is considered as promise."

This is based on the act of the Prophet (pbuh) himself who has executed conditional *hibah* as a promise. On another occasion, the Prophet (pbuh) had sent a souvenir to al-Najashi (King of Ethiopia), and he said to Ummu Salamah: "Indeed, I gave away to al-Najashi clothes and perfumes, but I think he had passed away. And I think that the souvenir will be sent back to me. If the souvenirs are returned to me, they will be yours."[6]

---

[5]Musnad Ahmad bin Hanbal, Bab Musnad Jabir bin Abdullah Radhiallahu anhu, Juz 3, p. 307, no. 14340.

[6]Musnad Ahmad bin Hanbal, Bab hadith Ummu Kalsum binti Uqbah Ummu Hamid, Juz 6, p. 404, no. 27317.

Ibn Qayyim holds that conditional *hibah* is valid based on the previous two *hadiths* (Ibnu Qayyim, 2004).

Secondly, the opinion of Ibn 'Abidin from the Hanafi school of thought who said (Ibnu 'Abidin, 2003):

امرأة تركت مهرها للزوج على أن يحج بها ، فلم يحج بها : قال محمد بن مقاتل : إنها تعود بمهرها ؛ لأن الرضا بالهبة كان بشرط العوض ، فإذا انعدم العوض انعدم الرضا ، والهبة لا تصح بدون الرضا.

"A woman gave up her dowry from her husband (as *hibah*), but with a condition (in return) her husband must perform pilgrimage (*Hajj*) together with her, but her husband did not fulfill the condition (perform *Hajj* together with her). Muhammad bin Muqatil said: "Indeed, she can claim her dowry, this is because her consent to make a gift is conditional to the consideration in the first place, when the consideration is not met, the consent is also absent, and such kind of *hibah* is considered void without consent."

Based on the above, Hanafis regard conditional *hibah* as binding upon the happening of the event. In this light, there is a legal maxim attributed to the Hanafis which goes as follows:[7]

"المواعيد بصورة التعاليق تكون لازمة"

"Promises in the form of ta'lik is binding" (Al-Nadawi, 2000).

Thirdly, according to the Maliki school of thought, *hibah* with conditions to be fulfilled in future is binding due to the reason that ambiguity (*gharar*) does not affect the validity of *hibah* contract. Ibn al-Rusyd mentioned that:

ولا خلاف في المذهب في جواز هبة المجهول والمعدوم المتوقع الوجود فتصح هبة الثمرة بعد بدو صلاحها

"There is no difference of opinions in Maliki school in permissibility of *hibah majhul* (giving away something unknown) and *hibah al-ma'dum* (giving away something that is not in existence) with the expectation to exist in the future. For this reason, it is valid to give away fruits which start to ripen." (Ibnu Rusyd, 1995).

As such, Malikis are also of the view that conditional *hibah* is binding upon the occurrence of a particular event.[8]

---

[7] Ali Ahmad al-Nadawi (2000). Vol. 1, p. 585.
[8] Al-Qarafi, *al-Furuq*, Beirut: Dar al-Kutub al-Ilmiyyah, no date, vol. 8, p. 28.

Second: The Opponents of Hibah Muallaqah

This is the opinion of Syafie jurists and some of the Hanbali and Zahiri jurists. According to Syafie school, *hibah* with any conditions in the future is not binding as *hibah* is a contract that will become invalid because of *jahalah* (ignorance). Consequently, *hibah* cannot be made conditional (to certain event in future) and this is similar to the prohibition of conditional sales contract (to certain occurrence in future) (Al-Syirazi, 1995).

Ibn Hazm, one of the prominent scholars of the Zahiri school of thought also said that: *"Hibah contract, in its nature, is a contract that cannot be made conditional"* (Ibnu Hazm, 1988).[9] In addition to that, Ibn Qudamah of Hanbali school mentioned:

"لا يصح تعليق الهبة بشرط : لأنها تمليك لمعين في الحياة ، فلم يجز تعليقها على شرط ، كالبيع"

> *"Hibah* contract which is conditional to terms is void because *hibah* is a giving of ownership (transfer of ownership) to a specific beneficiary in his life, thus, *hibah* cannot be conditional like sale contract (which cannot be conditional)" (Ibnu Qudamah, 2004).[10]

However, Al-Buhuti argued that in the above situation, *hibah* is valid and ownership is legally transferred from the benefactor to the beneficiary of *hibah*, but the condition is void (ineffective) (Al-Buhuti, 2000).[11]

Characteristics of *Hibah Mu'allaqah*

1. *Hibah mu'allaqah* is not considered a contract of exchange but is still in the realm of the contract of *tabarru'* (gratuity). Hence, in a situation where there is *gharar* (ambiguity), whether the beneficiary will obtain the *hibah* or not and whether the condition will take place or not, it is still tolerable. On the contrary, in a contract of exchange, if the subject matter is *gharar* (ambiguity), it shall render the contract null and void.

---

[9]Ibnu Hazm (1988). vol. 8, p. 59.
[10]Ibnu Qudamah (2004). p. 5/384.
[11]Al-Buhuti (2000). p. 2/434.

2. *Hibah mu'allaqah* is considered as binding when the conditions or terms are fulfilled as per the opinion of Hanafis and Malikis.

3. *Hibah mu'allaqah* is considered similar to a promise from one party. It is a promise to give away something, but not a promise to engage in a contract of exchange (sale contract) in future. Most of the contemporary scholars allow the practice of promise from one party (unilateral) and regard it as binding. This is in accordance with the resolution of Fiqh Academy (*Majma Fiqhi*) and AAOIFI standards. Nevertheless, they prohibit the practice of bilateral promises from both parties to engage in a *mu'awadah* contract in the future due to the fact that it is considered as similar to concluding a sales contract on a future basis. This practice is not permissible on the basis of concluding a sale contract by deferring the exchange of the two considerations (*ta'jil al-badalain*), but with immediate binding effect.

Suggested Guidelines for *Hibah Mu'allaqah*

1. *Hibah mu'allaqah* cannot be used for the purpose of contradicting Shariah (Al-Nadawi, 2000)[12] or prohibited transactions such as a deception that will lead to *riba* (interest). For example, one party said to another party: "If the profit of *mudharabah/wakalah bi al-istithmar* (agency of investment) is less from the expected amount, I promise to give *hibah* to make good the short of the exact amount."

2. Selling of *hibah mu'allaqah* to a third party is not permissible.

3. *Hibah mu'allaqah* shall not be contradictory to the substance of *hibah* (*muqtada hibah*) such as *tamlik 'ala wajh al-ta'bid* (transfer of ownership forever without time constraint). If there is a time constraint, it will be considered as a borrowing contract, and not as *hibah*.

4. *Hibah* can only be binding upon the deceased of the benefactor, and it will be considered as a will (*wasiyyat*). Nonetheless, the rule is that the estates of the deceased shall not be distributed

---

[12] *Ibid* at note 7.

to the heirs through an execution of the will, but rather it shall adhere to the rule of Islamic laws of inheritance (*farai'dh*).

Application of *Hibah Mu'allaqah* in *Takaful* Industry

Nomination of Takaful Beneficiaries

A question arises whether the *takaful* beneficiaries in family *takaful* are only entitled to the *takaful* benefits similar to *tarikah* (inheritance) which are subject to Islamic laws of *farai'dh*, or whether it is permissible to give the *takaful* benefits to a sole individual beneficiary as practiced in conventional insurance when the policyholders pass away.

To explain the concern of ownership over *takaful* benefit, there is a need to find out how ownership can be established from Shariah perspective. In Islam, there are a number of means to acquire a valid ownership and absolute possession, they are as follows:

1. The contract of exchange (*mu'awadah*) such as selling and buying and any other contract which involve the exchange of two considerations through a mutual consent between the contracting parties. This includes contract of *ijarah* i.e. hire of tangible assets or services by individuals.
2. Gratuity and unilateral contract (*tabarru'*) which only involves one party and the ownership is transferred with mutual consent without any compensation; i.e. will (*wasiyyat*), gift (*hibah*), and endowment (*wakaf*.)
3. Substitute or *khalafiyyah*, such as heritage, the penalty payment *(diyat)* and compensation fees.
4. Taking possession over the legally permissible natural resources such as fish in the sea or water in the river.
5. The growth or yields from the existing owned assets such as eggs, cow's milk, fruits, etc.

These categories prove that one person may have ownership or possession which can be obtained through various reasons and means as mentioned above, either with his own efforts or the effort of others.

As far as *takaful* is concerned, the ownership of the *takaful* benefits is regarded as payment not against a consideration via the contract of unilateral gratuitous contract (*tabarru'*). The ownership is established through a binding promise to give *hibah* (compensation) to the *takaful* beneficiaries, but it is conditional to the event of peril.

One may argue that it is established by way of a contract of exchange where the *takaful* participants pay the contribution as consideration for the *takaful* benefits for the *takaful* beneficiaries upon his death. If he does not pay the contribution, then there will be no *takaful* coverage and the *takaful* operator will never pay such *takaful* benefits. A question arises whether the *takaful* participants' efforts to subscribe the *takaful* certificate and pay the monthly contribution is sufficient to confirm that he has the right over the *takaful* benefit as consideration for his payment of the *takaful* premium, and hence the *takaful* benefits shall be treated as inheritance (*tarikah*).

Nonetheless, to meet the Shariah requirement in changing money with money at par to avoid *riba*, *takaful* benefits in family *takaful* cannot be regarded as consideration for the payment of premium. Although there is a commitment by the *takaful* participants through the *takaful* contract, it is actually an effort made by the deceased, and the *takaful* benefits (which are considered as compensation money) are considered as *tabarru'* from the *takaful* Risk Fund on behalf of the rest of the *takaful* participants. The *takaful* operator only acts as participants' agent to manage the *takaful* contribution, underwriting businesses, investing the fund and paying the benefits to the deceased's family in the event of death.

Contributions paid by the *takaful* participants are actually a mutual guarantee for assistance against perils which can happen to any participating member. In addition to that, the amount of contributions received by the *takaful* operator from the participants is not owned by the *takaful* operator, but rather owned by the fund on behalf of the participants.

Another important point to emphasize is the fact that *takaful* benefit is not yet established until the event of peril happens. This is different from the contract of exchange where the obligation, commitment and transfer of ownership takes place immediately upon execution of contract. In *takaful* industry, the financial responsibility is directly based on a promise to help one another as agreed in the contract. This is in line with the concept of *hibah mu'allaqah*. In other words, *tabarru'* fund that is managed by the *takaful* operator as an agent to the *takaful* participants is committed to pay the *takaful* benefits to the *takaful* beneficiaries after the death of the *takaful* participants. In this case, the concept of conditional *hibah* or *hibah mu'allaqah* can be applied. Therefore, the benefits payable to the nominated beneficiaries actually come from all *takaful* participants who mutually agreed to support each other by making a conditional *hibah*: "Our contribution which shall be treated as *takaful* benefit is conditional on the happening of the event of peril" or in other words, "If the unfortunate event happens, we will assist you by paying compensation to your nominated beneficiaries."

Contemporary Islamic Rulings (*Fatwa*) That Allow *Takaful* Benefits to be Given to the Sole Beneficiary

In general, it is submitted that *takaful* benefit is similar to inheritance and will be distributed according to rule of *faraidh*. Nevertheless, recently, there has been some new *fatwas* that allow the distribution of the *takaful* benefits to a sole individual beneficiary, similar to the practice of conventional insurance.

(1) *Fatwa Nadawat al-Barakah* for example, states as follows:

"It is permissible to distribute the (*takaful*) death benefit according to the law of mirath (Islamic law of succession), as it is also permissible to distribute the payment to particular individuals or parties as specified by the participant on the basis that the benefit is the contribution of other participants to the beneficiary as specified by the participant and not his estate."[13]

(2) Shariah Advisory Council of Bank Negara in its 34th meeting held on April 21, 2003 resolved:

i. *Takaful* benefit can be used as *hibah* since it is the right of the *takaful* participants. Therefore, the *takaful* participants should be

---

[13]Collection of al-Barakah Fatwas 1981–1997, p. 173.

allowed to exercise their rights according to their choice as long as it does not contradict with Shariah.

ii. The status of *hibah* in *takaful* plan does not change into will (*wasiyyat*) since this type of *hibah* is a conditional *hibah*, in which *hibah* is an offer to the recipient of *hibah* for only a specified period. In the context of *takaful*, the *takaful* benefit is both associated with the death of the *takaful* participant as well as maturity of the *takaful* certificate. If the *takaful* participant remains alive on maturity, the *takaful* benefit is owned by the participant, but if he dies within such period, then *hibah* shall be executed.

iii. A *takaful* participant has the right to revoke the *hibah* before the maturity date because conditional *hibah* is only deemed to be completed after delivery is made (*qabadh*).

iv. *Takaful* participant has the right to revoke the *hibah* to one party and transfer it to other parties or terminate the *takaful* participation if the recipient of *hibah* dies before maturity; and,

v. The *takaful* nomination form has to be standardized and must stipulate clearly the status of the nominee either as a beneficiary or an executor (*wasi*) or a trustee. Any matter concerning distribution of *takaful* benefit must be based on the contract and the implication of every contract being executed should be clearly explained to the *takaful* participants.

Shariah Issues in Giving the *Takaful* Benefit (*Hibah*) to a Sole Beneficiary

There are a number of unsettled issues over sole beneficiary, among others:

(1) How to make *hibah* of something which is not yet realized i.e. there will be no death benefit if the *takaful* participant is alive until the maturity of the *takaful* certificate?

(2) If it is a *hibah*, can the *takaful* participant retract the *hibah* because he/she takes the *takaful* plan for his/her own benefit upon maturity?

(3) In a valid *hibah*, the ownership is transferred to the recipient. What if the recipient dies? It becomes his/her estate. Can the recipient be replaced?

(4) If a husband is paying a *takaful* plan for his wife, can he be the recipient of the *takaful* benefit or can he only be a trustee/executor and as such, the *takaful* benefit should be treated as her estate which shall be distributed according to rules of *mirath* or *faraidh*?

(5) *Hibah* which is tied up with death is a will (*wasiyyat*). It is not allowed to make *wasiyyat* to the inheritor (*waris*). Another issue is that it may not serve the purpose of protecting a particular recipient.

(6) How far is the procedure recognized by law?

Recommendations and Suggestions

This chapter proposes a solution where *takaful* participants can give suggestions regarding the names that they want to put as beneficiaries, but conditional *hibah* (*hibah mu'allaqah*) comes from other *takaful* participants, in which all of them agreed upfront to make such a *hibah* contingent on the occurrence of the event of peril. This conditional *hibah* will be executed by the *takaful* operator as the appointed agent. If there is a death case, the participants will give the *takaful* benefit to the selected beneficiaries nominated by the participant/policyholder. The *ta'lik* (condition) here is that upon the participant/policyholder's death, the *takaful* benefit (compensation money) derived from the *tabarru'* fund will be paid to the definite beneficiaries only as recommended by the *takaful* participant.

The nominee that has been chosen by the *takaful* participant among the beneficiaries does not receive a *hibah* from the *takaful* participant himself, but from all *takaful* participants. To avoid confusion, *hibah mu'allaqah* is handled by the *takaful* operator as an agent to the other *takaful* participants via *tabarru'* fund. It is suggested to differentiate between *tabarru' fund* and personal investment fund. The proceeds from the personal investment fund should be treated as the estate of the deceased and should be subject to the rules of *al-faraid*.

The *hadith* narrated by Jabir as mentioned before can be used as evidence to support this suggestion. As mentioned in the *hadith*, the Prophet (pbuh) said to Jabir: "If the wealth from Bahrain arrives, I will give to you from it this amount, (and the Prophet repeated it three times)." Unfortunately, the properties did not reach Medina until the Prophet (pbuh) was demised. When the properties safely reached Medina in the time of caliph Abu Bakar (R.A), he asked those whom the Prophet made any promises or any debt obligation claims against him, to let him know. Hence, Jabir (R.A) came to see Abu Bakar (R.A) and said that: "The Prophet had made promises to me." As a result, Abu Bakar (R.A) calculated the properties three times before handing it over to Jabir (R.A).

In this *hadith*, Abu Bakar decided to fulfil the promise made by the Prophet (pbuh) which is making *hibah* from the wealth coming from Bahrain to Jabir. In this *hadith*, it is noted that the subject matter of *hibah* was only realized after the demise of the Prophet (pbuh). It is not considered as a will (*wasiyyah*) because the Prophet (pbuh) does not make it conditional to his death. The condition is if the wealth from Bahrain can be realised and brought to Medina. This exercise is only considered as a promise (similar to the case where the Prophet made a promise to Ummu Salamah to give some of the presents that has been returned by King Najasyi) and it is not the Prophet's will although those gifts were bought to Medina after his demise.

Muslim jurists hold that there are two interpretations for the above *hadith* regarding *mal Bahrain* (wealth from Bahrain):

First: there is a likelihood that this occasion falls under the purview of what is exclusively dedicated to the Prophet (pbuh) only. Hence, all the promises made by him should be honored and are compulsory to be fulfilled because every promise by the Prophet (pbuh) is true and he did not mention something unless it is really true. In this *hadith*, the promised goods which were treated as *hibah* were not in the possession of the Prophet (pbuh), but it belonged to *Baitulmal*. In this case, one can argue that based on the capacity of a prophet, it is similar when the Prophet (pbuh) declared that some of the properties which were owned by *Baitulmal* must be given to certain people as well. The reason is because every command coming from the Prophet (pbuh) is absolutely true and must be fulfilled.

Second: there is also a possibility that the *hadith* can also be applied to all. It means that the rules and laws taken based on this case have not been restricted to the promises made by the Prophet (pbuh) only, but are also appropriate to all Muslims in general.

By referring to the above discussion, since there is a possibility of two interpretations of the above *hadith*, it is suggested that every promise to give *hibah* for every similar case which is conditional in nature can be considered as binding as per the opinion of some Maliki and Hanafi jurists. But, it should be borne in mind that it should not be in the form of *wasiyyah* (distributing the estate prior to death by making it conditional to death) to the heirs as it is prohibited by the Prophet (pbuh).

Charging and Assignment of *Takaful* Certificate

Among the pertinent issues in the industry is taking insurance certificate as collateral for financing facilities. This can be for corporate financing, personal financing and all ranges of financings. The practice is the same for *takaful* certificate. In the event of default or in the event of loss, the financier will have the right of the assigned *takaful* benefit.

From Shariah perspective, most of the contemporary scholars allow to charge or assign *takaful* policies as security for the payment of outstanding debts. This is acceptable and practicable from the legal perspective. The Shariah issue is how to take as collateral (*rahn*) something which is not in existence. This raises the issue of uncertainty (*gharar*). However by regarding it as a supportive contract, and by regarding it as an undertaking and obligation of the *takaful* operator on behalf of other participants as stated in the *takaful* certificate, one can argue Shariah-wise it is permissible.

The weightage of the *takaful* policies lies in the undertaking to donate or conditional *hibah* made by the participants. It is also argued that participants can determine the beneficiary of the *takaful* program in which they participate. Even family *takaful* certificates have been widely used as assignment for the settlement of debts such as the debt of personal financings.

Conclusion

In conclusion, the concept of conditional *hibah* or *tabarru'* is technically used for *takaful*. As for the distribution of *takaful* benefit, conditional *hibah* is taken from *tabarru'* fund, through the management of the *takaful* operator on behalf of all *takaful* participants to give financial assistance in the form of *takaful* benefits to the beneficiaries. As for family *takaful*, the nominee can be the sole beneficiary if the benefits come solely from the *tabarru'* risk fund, or it can also be divided based on Islamic law of inheritance, but the money in the participant's personal investment account is subject to Islamic law of inheritance. The proceeds of *takaful* certificates can be taken as collateral for financing activities which includes construction, sale and purchase, and other financial obligations. They can be assigned to the financiers even though the policyholders already proposed their family members as nominees of the beneficiary upfront.

Legislation

Islamic Financial Service Act 2013

References

Ali Ahmad al-Nadawi (2000). Jamharat al-Qawa'id al-Fiqhiyyah fi al-Mu'amalat al-Maliyyah, Vol. 1, p. 585. Riyad: al-Rajhi Bank.

Al-Mardawi, Ali bin Sulaiman (1997). *al-Insaf fi Ma'rifat al-Rajih min al-Khilaf ala Mazhab al-Imam Ahmad bin Hanbal*. Beirut: Manshurat Muhammad 'Ali Baydun.

Al Syaibani, Ahmad bin Hanbal (2001). *Al-Musnad*. Cairo: Muassasah al-Risalah.

Al-Buhuti, Mansur bin Yunus (2000). *Syarh Muntaha al-Iradat*. Beirut: Dar al-Fikr.

Al-Mardawi and Ali bin Sulaiman (1997). *al-Inshof fi Makrifah al-Rajih min al-Khilaf ala Mazhab al-Imam Ahmad bin Hanbal*, Vol. 20, p. 391. Bayrut: Manshurat Muhammad 'Ali Baydun.

Al-Nadawi, Dr Ali Ahmad (2000). *Jamharat al-Qawa'id al-Fiqhiyyah fi al-Mu'amalat al-Maliyyah*. Riyadh: al-Rajhi Foundation.

Al-Qarafi (n.d.). *Al-Furuq*, Vol. 8, p. 28. Beirut: Dar al-Kutub al-Ilmiyyah.

Al-Sharbini (2000). *Mughni al-Muhtaj*, Vol. 2, p. 396. Beirut: Dar al-Fikr.

Al-Syirazi, Abu Ishaq Ibrahim bin Ali bin Yusuf (1995). *Al-Muhazzab fi Fiqh al-Imam al-Syafie*, p. 2/332. Beirut: Darul Kutub al-Ilmiyyah.

Ibnu Rusyd, Abu al-Walid Muhammad bin Ahmad (1995). *Bidayah al-Mujtahid wa Nihayat al-Muqtasid*, p. 2/329. Cairo: Darul Salam.

Ibnu 'Abidin, Muhammad Amin (2003). *Hashiyat Ibnu 'Abidin : Rad al-Muhtar Ala al-Dur al-Mukhtar*, p. 5/710. Riyadh: Dar A'lim al-Kutub.

Ibnu Hazm, Ali bin Ahmad (1988). *Al-Muhalla bi al-Athar*. Beirut: Dar al-Kutub al-'Ilmiyyah.

Ibnu Qayyim al-Jauziyyah, Shamsuddin Muhammad bin Abi Bakr (2004). *Ighasat al-Lahfan min Masoyid al-Syaitan*, p. 2/16?17. Cairo: Dar Ibn al-Haytham.

Ibnu Qudamah, Mauqifuddin Abdullah bin Ahmad (2004). *Al-Mughni*. Cairo: Dar al-Hadith.

Chapter 16

Managing the Risk of Insolvency in the Construction Industry with Equity Financing: Lessons from *Nakheel*

ABDUL KARIM Abdullah

Introduction

The concept of risk has attracted much attention over the course of history, due to its volatile character. Human beings can be generally expected to act rationally. However, even the most "rational" decisions do not always guarantee a desired outcome, as not all relevant factors may have been taken into account when a specific decision was arrived at. Thus, even the most "rational" decisions may not take into account unexpected events.

In other words, historical experience confirms that there may always be a variable that may have been overlooked or one that was erroneously weighed. The occurrence of disaster in any form, due to the failure to recognize, assess and reduce risk, has caused much hardship, both in the form of loss of life or damage to property.

A major risk faced by businesses is the risk of bankruptcy. This risk arises from the uncertainty inherent in the business environment and a failure to take pro-active measures to minimize risk. It arises especially in relation to corporations that opt for financing investments by borrowing rather than issuing share capital or, in the case of Islamic finance, participatory, asset-backed *sukuk* that enable counterparties to share risk.

This chapter thus argues that, to reduce the risk of insolvency, it is better to finance investment by selling shares rather than borrowing. The reason is that equity financing or its equivalent in Islamic finance — asset-backed *sukuk* — does not oblige companies to commit themselves to making predetermined payments, regardless of the prevailing business environment.

In good times, debt financing normally does not present a problem. However, during economic downturns, loan financing can impose a great burden on the indebted company. When cash flows of an enterprise turn out to be lower than expected, default becomes a real possibility. When profits fall below the level necessary to service loans, lenders may force defaulting businesses into bankruptcy.

By contrast, this cannot happen to businesses funded by means of equity financing. Such businesses remain debt-free and therefore should be able to survive periods of low profits or even losses. This means that the long-term viability of enterprises financed by means of equity are better than those financed by loans: "the probability of bankruptcy and financial distress are increased when debt-based financing is used." (Arshad and Asad, 2001).

This is in sharp contrast to issuers of shares, who are under no legal obligation to pay profits to shareholders (dividends) when no profits have been earned.[1] Thus, financing investment by selling shares is more favorable to the success of the firm in the long term, as equity financing does not require companies to pay profits to shareholders regardless of the market environment.

Advantages of Equity Financing

Equity financing does not require businesses to go into debt. Remaining debt-free protects businesses from the risk of default. Equity financing is "less risky in terms of cash flow commitments.

---

[1] "In bad times, interest payments must continue at the same rate, while equity-based payments are reduced." *Loc. cit.*

There is no obligation or liability to distribute or service profit when there is no profit made."[2] There may be losses from time to time, but there can be no defaults.

By contrast, borrowing presents the risk that companies may take on *excessive debt*. The result may be overinvestment, a high degree of indebtedness and perhaps defaults, arising from the failure to sell or lease infrastructure within a given time period. This would be particularly true in instances where plans are based on unrealistic expectations.

It may be objected that equity financing is harder to get, as investors prefer the assurance of interest and capital guarantees. However, business ventures with convincing prospects of success should face little difficulty in attracting sufficient funding. Business proposals with poor prospects, by contrast, will attract few investors.

That borrowing is risky has been amply demonstrated in the recent (2007) financial crisis. Financial institutions borrowed heavily and purchased risky debt securities that offered — at least for a time — higher returns. This resulted in high debt-to-equity ratios. The collateralized debt obligations (CDOs), purchased with borrowed funds were assigned AAA (investment grade) ratings, despite the fact that much of the collateral backing the CDOs consisted of subprime mortgages, an uncertain source of revenue under any condition.

Sharp rises in rates in the years prior to 2007 triggered extensive defaults, first in the adjustable rate sub-prime mortgages. Due to impaired cash flows from borrowers to creditors, rating agencies downgraded the CDOs. The downgrades triggered dramatic declines in their prices. Financial institutions were unable to sell the CDOs at prices that would enable them to repay the money they borrowed to buy them in the first place.

In retrospect, the higher yields initially paid to the buyers of CDOs were dwarfed by the steep falls in their market values.

---

[2]n.a. (2009) *The Islamic Securities (Sukuk) Market*, Securities Commission Malaysia, LexisNexis, p. 14.

Investment banks such as Lehman Brothers went bankrupt. Others nearly went bankrupt. The prospect of a systemic collapse became real. Only massive infusions of funds (bailouts) by central banks averted a domino-like financial collapse of institutions that have been allowed to become "too big to fail" due to poor or nonexistent regulatory oversight.

In the meantime, the progressively larger bailouts of private institutions using public funds[3] sowed the seeds of the next round of inflation and the next, still bigger bubble, in a familiar cycle of boom and bust.

Losses by insolvent companies have been estimated in the trillions of US dollars.[4] In light of some defaults and near defaults, misgivings have emerged about the benefits and even the long-term viability of loan financing *vis-à-vis* equity financing.

Risk is a powerful deterrent. Investors are motivated by the hope of gaining profits and restrained by the risk of suffering losses (Chapra, 1985). The possibility (risk) of suffering losses acts as a powerful incentive to investors to allocate resources wisely. Where investors feel there is little or no risk, they may be expected to be inclined to allocate resources on a greater scale than what is warranted by a better assessment of risks and returns. Risk constitutes an important incentive for exercising due diligence, a *sine qua non* of an efficient allocation of capital.

---

[3] This process is known as the "socialization of losses," as distinguished from the "privatization of profits."

[4] According to the USA Congressional Service, the cost of the bailout resulting from the 2007 global financial crisis amounted to USD 8.1 trillion, measured in today's prices. This exceeded the combined costs of all economic crises in USA history. These included the following, in descending order: World War II at USD 4.1 trillion, Vietnam War at USD 686 billion, Iraq War at USD 648 billion and counting, the New Deal at USD 500 billion, Korean War at USD 320 billion, World War I at USD 253 billion, War on Terror at USD 171 billion and counting, and Marshall Plan at USD 115 billion. Comment by kenj, added on December 30, 2010, in McConnell, P (2010). Regulatory madness in the banking world. ABC, *The Drum Unleashed*. https://www.abc.net.au/news/2010-12-21/regulatory_ma dness_in_the_banking_world/42466 [Accessed November 7, 2018]. Dr. Patrick McConnell is a Visiting Fellow at Macquarie University Applied Finance Centre, Sydney, Australia.

At the macroeconomic level, a major result of the underutilization of equity financing in the 2008 global financial downturn was a misallocation of resources in the housing industry on a massive scale. This was evident in the millions of houses that were constructed in the USA and financed by risky subprime mortgages. Most of these houses have been repossessed by financial institutions and have been torn down due to dilapidation. Trillions of dollars were wasted.

The need to face (share) risk filters out excessively risky investments. Ill-conceived projects are unlikely to attract funding. The fact that a given project is unable to attract funds shows that it should not be financed in the first place. As a consequence, "white elephant" projects will have difficulty attracting sufficient financing, thereby saving valuable resources for more beneficial uses elsewhere.

Nakheel Construction Company

Between January 2006 and December 2007, before the economic and financial crisis of 2008, Dubai companies borrowed heavily using both conventional as well as Islamic bonds.[5] In January 2006, Ports, Customs and Free Zone Corporation (PCFC) of Dubai issued USD 3.52 billion convertible *musharakah* sukuk, guaranteed by Dubai World, to fund DP World's acquisition of P & O, the UK's biggest ports and ferries operator, and the third largest in the world (Muhammad Ayub, 2007). Dubai World in turn purchased the Turnberry golf course and a 21% interest in the London Stock Exchange. In December 2006, Nakheel World, a subsidiary of Dubai World, issued USD 3.5 billion sukuk. An additional USD 750 million issue followed. Jebel Ali Free Zone issued USD 2 billion and Dubai World USD 1.5 billion worth of sukuk (Lahem, 2010).

---

[5]The vast majority (85%) of the so-called "Islamic bonds" in effect amounted to "replicas" of conventional bonds, insofar as they required issuers to guarantee both periodic payments and the principal amount invested to the providers of the funds. Thus, in substance little difference existed between Islamic bonds structured this way and conventional bonds.

In 2007, Dubai World purchased the luxury liner QE2. In October 2008, on the eve of the global financial crisis, an ambitious USD 38 billion development plan, which included the tallest tower in the world, was announced. Dubai World and its investment arm, Istithmar purchased a number of "trophy properties." The purchases included:

> CityCenter Casino & Resort, a large Las Vegas development in which Dubai World teamed with MGM Mirage. Dubai World's share of the CityCenter investment was USD 5.4 billion... ; Barneys, the luxury retailer, bought in 2007 for USD 942 million; 450 Lexington Ave.... in Manhattan, bought for USD 600 million in 2006; a stake in the Mandarin Oriental, a 248-room hotel in Manhattan ... in 2007... at USD 380 million; and a 50% stake in the Fontainebleau Miami, an 876-room resort hotel in Miami... for USD 750 million; [and] the iconic art-deco former Adelphi hotel building on the Strand, WC2 (Jones and Petterson, 2009).

By the end of 2009, Dubai World and its companies accumulated debt of over USD 100 billion to over 100 lenders.[6] Its debts at the time exceeded its GDP (Mortished, 2009a). USD 50 billion of its liabilities were scheduled to mature by 2013 (Balzli *et al.*, 2009). Publicly, Dubai acknowledged USD 80 billion in liabilities.[7]

The credit squeeze and a 50% drop in oil prices in 2008 from their peak levels ended a six-year boom, fuelled mainly by high oil prices. Sales, profits and asset prices declined. Property prices dropped 50% from their pre-crisis peak in 2008. In Saudi Arabia the stock market lost up to two thirds of its market capitalization (Allam, 2010). "Cheap money, leverage and expectations of ever-rising property prices generate 'hot money inflows' which ultimately reverse in spectacular fashion when the bubble bursts" (Mackinnon, 2010).

The Islamic bond market "has not escaped the throes of the credit crisis... with investment banks and finance houses worldwide still reeling from the collapse of the USA sub-prime mortgage market and the breakdown of the wholesale money markets amid

---

[6]Dredging the debt (October 29, 2009). *The Economist.* http://www.economist.com/node/14753834 [Accessed September 19, 2010].

[7]A financial sandstorm (November 30, 2009). *The Economist.* http://www.economist.com/node/15004072 [Accessed September 19, 2010].

persistent counterparty risk concerns and deep-seated investor distrust in credit-sensitive assets" (Hesse *et al.*, 2009). Poor investment decisions contributed to problems.[8]

With the onset of the economic and financial crisis in 2008, the market for luxurious properties weakened. Sales and rentals of properties slowed. There was not enough demand for the newly constructed facilities to generate the income necessary to pay off all the commitments on a timely basis.

Real estate prices have collapsed and are now only half of what they were a year ago. And yet new villas and luxury condominiums are still being completed everyday — and are standing empty. Entire floors are deserted in the skyscrapers along Sheikh Zayed Road, and giant banners with the words "To Let" are displayed across the fronts of buildings. Real estate experts estimate that only 41% of office space is occupied, and the vacancy rate is only expected to increase. In 2011, Dubai, a city of about 1.5 million citizens, will have more office space available than Shanghai, which is 10 times as large as Dubai (Balzli *et al.*, 2009).

Since the crisis started, approximately USD 430 billion of planned development projects, the majority of them in Dubai, have been cancelled (Cummins *et al.*, 2009). On February 22, 2009 the Abu Dhabi-based central bank of the United Arab Emirates bought USD 10 billion of *sukuk* from Dubai (Anwar, 2010). Two banks backed by the government of Abu Dhabi bought another USD 5 billion of Dubai World bonds.[9] In October 2009, one month before the call

---

[8]Development in Dubai included the construction of the world's tallest building and 40,000 yet-to-be completed luxury (up to 7,000 square feet) homes on three man-made islands, shaped like palm trees. It also includes the construction of more than 100 luxury hotels, inclusive of several seven star hotels, and a ski-slope in the Mall of the Emirates. Nakheel, the subsidiary of Dubai World built *The World*, a cluster of 300 islands shaped like the world, offshore of Dubai. Cummins, C, S Bianchi and M Sleiman (2009). Dubai: A high rise, then a steep fall. *Wall Street Journal*. http://online.wsj.com/article/SB125988807548075805.html?KEYWORDS=sukuk [Accessed October 6, 2010].

[9]Abu Dhabi, the chief of the Emirates, controls 7% of global and 90% of the Emirates' oil resources.

for a standstill on *sukuk* issued by Nakheel, Dubai World sold an additional USD 2.5 billion of *sukuk*. In the first half of 2009 Nakheel declared a loss of USD 3.65 billion. This was compared to a profit of USD 722 million during the same period a year earlier. Losses were attributed to slower sales and write-downs in the values of properties (Moya, 2009).

With the global economic downturn in full swing, sales and rentals of newly constructed waterfront residential and commercial properties did not generate enough revenue to redeem USD 4.1 billion worth of *sukuk* — the largest ever issued — which matured on December 14, 2009. On November 25, 2009, less than three weeks before the maturity date, Dubai World announced — in a 200-word statement — that it was seeking six additional months to repay USD 26 billion of maturing liabilities, including the USD 4.1 billion Nakheel *sukuk* due on December 14, 2009. The announcement sent shockwaves through the global financial markets, raising fears of default on sovereign debt (Sakoui, 2010). It also triggered an extensive capital flight. "Credit-rating agencies quickly downgraded all government-related debt."[10]

The subscribers (buyers) of the Nakheel *sukuk* issues comprised foreign and local investors. A refusal to grant a standstill or restructuring would trigger default and formal bankruptcy proceedings. The latter would have been complex due to the fact that a number of the *sukuk* agreements were rather opaque regarding the rights and responsibilities of the counterparties.

On November 30, 2009, the government of Dubai announced that it *did not guarantee* the debts of Dubai World. This prompted fears that its creditors could lose billions of dollars. Many investors assumed that the debts were "backed by the government of Dubai, and ultimately by Dubai's oil-rich neighbour, Abu Dhabi."[11]

---

[10]Standing still but still standing (November 26, 2009). *The Economist*. http://www.economist.com/node/14977157 [Accessed September 19, 2010].

[11]Default lines: What would happen if a member of the euro area could no longer finance its debt? (December 3, 2009). *The Economist*. http://www.economist.com/node/15016124 [Accessed September 19, 2010].

In a television interview on the same day, the director-general of the Department of Finance, Abdul Rahman al-Saleh, announced that lenders to Dubai World were in for "short-term pain." "Creditors need to take part of the responsibility for their decision," he stated. "The government is the owner of the company, but since its foundation it was established that the company is not guaranteed by the government" (Arnott, 2009). These announcements "reminded investors that tacit sovereign guarantees may be worthless."[12] Nakheel *sukuk* promptly fell to 38 cents on the dollar (Thomson, 2010). As a result, "fears surfaced that *sukuk* failed to provide the same legal protection as conventional bonds... The concern is that *sukuk* creditors may not be protected" (Mortished, 2009b). The "uncertainty about investor protection" damaged investor confidence (Hesse and Jobst, 2010). Holders of *sukuk* realised they had a "limited ability to lay their hands on their assets" (Usman, 2010). There was doubt as to whether the *sukuk* holders would be legally able to take possession of assets underlying the securities in a time of distress.

On December 14, 2009, just hours before USD 4.1 billion of *sukuk* was due, Abu Dhabi announced a USD 10 billion loan to Dubai (Low and Critchlow, 2009). In March 2010, Dubai presented a restructuring plan of USD 9.5 billion. On May 20, 2010 Dubai World announced that it reached agreement with a group of banks to restructure USD 23.5 billion of loans.

The 97 creditor banks of Dubai World include four British banks, Hongkong and Shanghai Banking Corporation Limited (HSBC), Lloyds, Royal Bank of Scotland (RBS), and Standard Chartered. Others include the Bank of Tokyo-Mitsubishi, Abu Dhabi Commercial Bank and Dubai's Emirates NBD bank. These seven banks hold 60% or USD 14.1 billion of Dubai World's debt (Reuters, 2010). "Lenders will wait up to eight years to get their USD 14.4 billion back but have avoided a 'haircut' on their principal under the terms of the deal, which offers 1% cash interest and an

---

[12]Ibid at note 7.

extra 1.5–2.5% per annum rolled up into a lump sum payment on maturity."[13]

Conclusions and Recommendations

During the recent economic crisis, stock and property markets fell in the Gulf countries just as they had declined in the rest of the world. All markets experienced reduced growth, lower asset prices and tighter liquidity.

Issuing debt or quasi-debt (Islamic bonds) exposed Nakheel to the risk that faces all companies that utilize debt in preference to equity financing, in particular the risk of default.

Issuers of *bona fide* profit and loss-sharing securities do not face risks of default and bankruptcy. When losses are experienced, the management declares a "loss," and no dividends are paid. Shareholders may, however, recover their funds by selling their shares in the stock market.

It goes without saying that the price they may be able to get in the share market depends on the performance of the business. When the business performs well, investors are likely to obtain higher prices and realize a capital gain. Where the business does not perform up to expectations, investors may be forced to accept less than what they paid for the shares in the first place.

The possibility that equity instruments such as shares may fail to pay dividends from time to time should not be cause for concern, as the possibility of incurring losses is balanced by the prospect of earning profits, especially over the longer term.[14]

References

A Financial sandstorm (November 30, 2009). *The Economist*. http://www.economist.com/node/15004072 [Accessed September 19, 2010].

---

[13]Dubai World clinches $23.5 bn debt deal (May 20, 2010). *The National*. https://www.thenational.ae/business/dubai-world-clinches-23-5bn-debt-deal-1.575888?videoId=5771275459001 [Accessed November 7, 2018]. (May 21, 2010).
[14]In a sound business venture, total profits will generally be higher over the longer term than total losses, thus enabling the company to grow and prosper.

Allam, A (2010). Al Rajhi turns to untapped sukuk. *Financial Times*. http://www. ft.com/cms/s/0/77d7767a-5714-11df-aaff-00144feab49a.html [Accessed July 21, 2010].

Anwar, H (2010). Dubai bonds to rally as default risk wanes, Deutsche Bank says. *Bloomberg Businessweek*. https://www.arabianbusiness.com/dubai-bonds-ral ly-as-default-risk-wanes-292877.html [Accessed November 7, 2018].

Arnott, S (2009). Gulf state turns its back on Dubai World's $59 billion debt. *The Independent*. http://www.independent.co.uk/news/business/news/gulf-state-turns-its-back-on-dubai-worlds-59bn-debt-1831758.html [Accessed August 10, 2010].

Arshad, Z and Z Asad (2001). Interest and the modern economy. *Islamic Economic Studies*, 8(2), 65.

Balzli, B, A Jung and B Zand (2009). Dubai's debt woes unsettle financial world. *Spiegel Online International*. http://www.spiegel.de/international/world/0,1 518,664225-2,00.html [Accessed September 17, 2010].

Chapra, MU (1985). *Towards a Just Monetary System*, p. 69. The Islamic Foundation.

Cummins, C, S Bianchi and M Sleiman (2009). Dubai: A high rise, then a steep fall. *Wall Street Journal*. http://online.wsj.com/article/SB125988807548075805 .html?KEYWORDS=sukuk [Accessed October 6, 2010].

Default lines: What would happen if a member of the euro area could no longer finance its debt? (December 3, 2009). *The Economist*. http://www.economist.c om/node/15016124 [Accessed September 19, 2010].

Dredging the debt (October 29, 2009). *The Economist*. http://www.economist.com /node/14753834 [Accessed on September 19, 2010].

Dubai World in $23.5 debt deal with banks (May 21, 2010). *Arab News.Com*. http:/ /arabnews.com/economy/article55913.ece [Accessed September 13, 2010].

Hesse, H and A Jobst (2010). Debriefing Nakheel — Wider implications for the Sukuk market. *Roubini Global Economics*. http://www.roubini.com/emerging markets-monitor/258811/debriefing_nakheel_-_wider_implications_for_the_ sukuk_market [Accessed October 9, 2010].

Hesse, H, A Jobst and JA Sole (2009). Islamic securitization — The right way forward?. *Roubini Global Economics*. http://www.roubini.com/financemarkets-monitor/255550/islamic_securitization_-_the_right_way_forward [Accessed October 9, 2010].

Jones, DH and AV Petersen (2009). A Dubai world debt and Nakheel Sukuk — Apocalypse Now? Distressed Real Estate Alert. *K & L Gates*. http://www.klgates.com/a-dubai-world-debt-and-nakheel-sukuk--a pocalypse-now-12-10-2009/ [Accessed November 7, 2018].

Lahem, al-Nasser. Is Sukuk fever over? (April 17, 2010). *Asharq Alawsat*. https://e ng-archive.aawsat.com/theaawsat/business/is-sukuk-fever-over [Accessed November 7, 2018].

Low, CB and A Critchlow (2009). Dubai World promises but the damage may be done. *Wall Street Journal*. https://www.wsj.com/articles/SB100014240527487 03954904574596013729588066 [Accessed November 7, 2018].

McConnell, P (2010). Regulatory madness in the banking world. *ABC, The Drum Unleashed*. http://www.abc.net.au/unleashed/42466.html [Accessed May 3, 2011].

Mackinnon, N (2010). Dubai debt woes give Islamic finance its first big crisis. *asiaoneNews*. http://www.asiaone.com/News/Latest+News/Business/Stor y/A1Story20091202-183596.html [Accessed October 13, 2010].

Mortished, C (2009a). Nakheel debt talks turn spotlight on future of the Islamic bond market. *The Times,* http://business.timesonline.co.uk/tol/business /industry_sectors/construction_and_property/article6913156.ece [Accessed October 7, 2010].

Mortished, C (2009b). Western investors watch nervously as worth of Islamic bond is tested. *The Times.* http://business.timesonline.co.uk/tol/business/markets /the_gulf/article6934074.ece [Accessed October 7, 2010].

Moya, E (2009). Six Dubai companies downgraded to junk status. *The Guardian.* http://www.guardian.co.uk/business/2009/dec/08/dubai-comp anies-downgraded-rating-action [Accessed August 10, 2010].

Muhammad Ayub (2007). *Understanding Islamic Finance*, p. 389. John Wiley and Sons Ltd.

n.a. (2009). *The Islamic Securities (Sukuk) Market.* Securities Commission Malaysia, p. 14. LexisNexis.

Reuters (2010). Dubai may launch dollar Sukuk this year — Bankers. *Asharq Alawsat.* http://www.tradearabia.com/news/bank_181465.html [Accessed November 7, 2018].

Sakoui, A (2010). Sukuk: Sustained recovery expected in second half of the year. *Financial Times,* http://www.ft.com/cms/s/0/3c123094-5c91-11df-bb38-00144feab49a.html [Accessed July 21, 2010].

Standing still but still standing (November 26, 2009). *The Economist.* http://www.economist.com/node/14977157 [Accessed September 19, 2010].

Thomson, A (2010). The Calculus behind Dubai's Nakheel's Repayment. *Wall Street Journal.* https://www.wsj.com/articles/SB1000142405270230342980457514984 2965628952 [Accessed November 7, 2018].

Usman, H (2010). Islamic Finance's sukuk explained. *Financial Times.* http:/www. ft.com/cms/s/0/cec38bf2-440b-11df-9235-00144feab49a.html [Accessed July 9, 2010].

Index

4 Ps of marketing, 74
5 Ps model, 75

'Aqd, 29, 30, 85
'Aqidah, 8, 11
'Aqilah, 259
'ain, 177
'aqilah, 258
acceptance, 31
accountability, 57
ad-deen, xxi
Additional Dispute Resolution, 146
adjudication, 156–158
    statutory adjudication, 158
'adl, 174
ADR, 146–148
affordable housing, 205–207, 213
Agency (wakalah) model, 262
agents, 132
akhirah, 59
Al-Amanat, 37
al-Bai Bithaman Ajil, 86, 182, 187
Al-Ibra', 36
al-ihtikar, 85
    Ihtikar, xxiii
Al-ijarah, 49, 84, 87
    Ijarah, 35, 189
al-Istisna', 49
    Istisna', 90–93, 95, 96, 107, 108,
    111–114, 116, 117, 121, 123, 125,
    127, 128, 175
al-masālih al-mursalah, 20

Al-Muamalat, xx, 71, 84, 85, 90
    Muamalat, 84, 253
Al-Mudharabah, 86
Al-Muqawalah, 87, 91
Al-Murabahah, 87
    Murabahah, 172, 182
Al-Musharakah, 87
    Musharakah, 35, 189, 246
Alternative Dispute Resolution
    (ADR), 146
Appropriated Dispute Resolution, 146
aqīdah, 64
aqd, 49
Aqidan, 85
arbitration, 146, 148–152, 154
arbitrator, 50
arkan, 30
as-Sunnah, xxi
    Sunnah, xxiii, 19, 20
Assisted Dispute Resolution, 146

Bai al-Istisna', xxiv, 34
Bai al-Salam, 33
Bai al-Wafa', 34
Bai Istijrar, 34
Bai Muajjal, 33
Bai Muzayadah, 35
Bills of Quantities (BQ), 90, 109,
    121–125, 127
British East India Company, 129
Building Information Modelling
    (BIM), 123

Caliphate, 56
Civil Engineering Standard Method of
    Measurement (CESMM) for civil
    engineering works, 124
commenda, 132, 133
common law, 131, 133
common law rule, 130
compensation, 56
compound interest, 183
construction, xxiv, 29, 88
        construction industry, 49, 145
construction marketing, 73, 75, 76
consultants, 109
contract, 30, 49, 83, 88, 107, 233
        conditions of contract, 32, 233
        construction contract, 72, 83, 88,
        139
        construction works contracts,
        109, 121
        contract documents, 89
        conventional contracts, 108
        conventional-styled contract, 90
        Ethical marketing, 73
        Shariah contracts, 29
        Shariah-compliant construction
        contract, xxiv
        Shariah-compliant contract, 90,
        182
        Types of Contracts, 32
contract law, 132
contractor, 88, 107–109, 112, 117, 140
contractual relationships, 111
conventional finance, 173
cost plus, 90
cost plus/reimbursable contracts, 109
Creed of Absolute Monotheism, 7
Crowdfunding, 195

Darūriyyah, 18
Defects Liability, 141
Design Liability, 140
Design-Build-Finance-Operate-
    Maintain (DBFOM),
    214
Development, 40

Disputes, 145, 146
    Construction disputes, 145, 154
dividend, 173
Drawings, 122, 127
drawings and specifications, 90, 109

Employer, 107–109, 112, 117, 140
Endowment (*wakaf*) model, 263
English Commercial law, 129
English law, 129
equity, 135
equity financing, 288, 289
Equity-Type Home Ownership
    Scheme, 195
European legal system, 131
Ex-post equity condition, 242
Expected utility theory, 228

Fard, xxiii
fatwa, 43, 211
fiqh, 44
full recourse finance, 174, 176

Gharar, xxiii, 23–26, 85, 223, 224,
    225–229, 233, 234, 260
governance, 61, 63

Hājiyyah, 18
Habl habla, 225
halal, 176
halalan toyyiban, xxiv
Haram, xxiii, 176
Hasat, 226
hibah, 36, 272–275, 280
        Hibah mu'allaqah, 272, 275, 276
Hisbah, 61, 62
home financing, 185
        BBA Home Financing, 185
hudud, 10, 46
Hybrid model, 263

Ihsan, 65
Ihtikār, 28
ijab, 85
ijara, 172
        Ijara mawsufah fi al-dimmah, 175

ijma', xxi, 20
Ijtihad, xxi
Iktinaz, xxiii, 84
inaccessibility, 205
inflation, 183
insurance, 141
    conventional insurance, 245, 257
interest rate, 169
Investors, 290
Islam, xxi, 46, 47, 57
Islamic belief, xx, 71, 84
Islamic finance, 169, 172, 173, 223, 288
Islamic financial institution, 112, 113,
    115, 119
Islamic financial products, 170
Islamic full-recourse finance, 178
Islamic law, 10, 19, 43, 44, 46, 52, 61, 62
Islamic Project Finance, 170, 175
Istibdal, 217
istihsān, 20
istishāb, 20
Istisna', 87, 110, 115, 118, 119, 122
Itqan, 65, 66

jahalah, 275
joint venture, 212–214
Ju'alah, 37, 87

Kafalah, 36
khilafa, 59
Khilafah, 42
kifalah, 63

limited liability, 127–135, 139
    limited Liability Act, 131
    unlimited liability, 135
limited-recourse, 174
litigation, 145, 146
lost camel, 226
lose-lose, 242
lump sum, 90, 109

Ma'aqud Alaid, 85
Makruh, xxiii
Manipulation, 28

Maqasid al-Shariah, xxi, xxiii, 8, 12, 17,
    18, 38, 171, 177, 186
Mardatillah, 41
Marketing, 72–74
    Marketing practice, 72
Marketing Mix, 74
Maslahah, 84, 199
Maysir, 27, 28, 85, 226, 260
mediation, 148, 152–155
monetary system, 184
Mubah, xxiii
Mudarabah, 35, 138, 171
Muhtasib, 62
Mulāmasa, 225
Munābadha, 225
Murabaha Syndrome, 173
Murabahah Sale, 33
Musharakah Mutanaqisah, 182, 189,
    190

Najsh, 28, 29
nass, 84
niyyah, 45
non-recourse, 174, 177

objects of contract, 31
obligation, 139
offer, 30

parallel istisna', 113–115
peculium, 132
people, 75
performance bond, 142
pillars of Islam, 11
Place, 75
Price, 75
price of a contract, 89
principal, 132
procurement, 88, 125
product, 75
profit-loss sharing system, 241
profit-rate, 173
Profit-sharing (*mudharabah*) model, 262
project finance, 173, 174
Project financing, 170
project funding, 214

project management, 58
promotion, 75

qabul, 30
qisas, 10, 46
Qiyās, 20
quantities, 124
quantity Surveyors, 125
Quran, xxi, 19, 20
qurud hasanah, 135

Rahn, 36
real economy, 184
Regret Theory, 226, 228, 229
reimbursable, 90
retention fund, 142
riba, xxiii, 20, 85, 169, 174, 178, 260
    non-riba, 177
    riba-free, 170
ridha, 31
risk, 235, 254, 287, 289, 290
    risk avoidance, 255
    risk reduction, 255
    risk retention, 255
    risk transfer, 255
    risk-sharing, 176, 224, 254, 255
Rububiyyah, 42

sad al-dhara'i, 20
sadaqah, 64
Salam, 87
salat, 64
Shariah, xx–xxiii, 3–5, 7–10, 15–17, 20,
    38, 44, 56, 60, 71, 83, 160, 172, 177,
    223, 224
    Shariah compliance, 107
    Shariah guidelines, 41
    Shariah incompliance, 224
    Shariah law, 50, 51
    Shariah Principles, 50
    Shariah-compliant, xx, xxii–xxiv,
        83, 107, 110, 111, 114, 115,
        117–119, 176, 177, 218, 241, 248,
        271
    Shariah-compliant construction
        contract, 83, 86, 92

Shariah-compliant construction
    marketing, 76, 77
Shariah-compliant project
    financing, 122
Shariah-compliant project
    funding, 118
sharikah, 35, 135, 136
    sharikah al 'aqd, 136, 137
    sharikah al mulk, 136
Shirkah, 35
Shirkat-al-Milik, 189
shura, 64, 65
Sighah, 85
Special Purpose Vehicle (SPV), 214
Specifications, 122, 127
SPV, 215, 216
standard forms of contract, 90
Standard Method of Measurement
    (SMM) of Building Works, 124
Straits Settlements, 129
Sub-Game Perfect Equilibrium, 236
Subject Matter, 31
Sukuk, xxiv, 87
Sulh, 37, 158–162
sustainable development, 63

ta'awun, 256
ta'zir, 10
tabarru', 256
Tadlīs, 28, 29
Tahkim, 159–162
Tahsīniyyah, 19
takaful, xxiv, 87, 247, 254, 255, 257, 258,
    260–263, 265, 271, 277–279
    Family takaful, 263
    General takaful, 263
talaqqi al-rukban, 84
Tawhid, 10, 11, 42, 59
Tazkiyah, 42
tendering, 88
traditional dispute resolution, 145

'Uqud Al-Mu'awadat, 32
'urf, 20
ummah, 249
uncertainty, 223

unlimited liability, 130
Usul fiqh, 54

Virtual Reality, 123

Wajib, xxiii
Wakaf, 36, 64, 205, 207, 208, 210, 211, 213–217

Wakalah, 37
win-win, 242
Win-Win/Lose-Lose, 232
win-lose, 227
Wasatiah, 53

zakat, 63, 64, 205, 209–211, 213–218
zero-sum game, 226

Note: Words with more than one spelling

No	Word	Other Spelling
1	al-ihtikar*	Ihtikār
		Ihtikar
2	Al-ijarah*	Ijarah
		Ijara
3	al-mu'amalat*	Al-Muamalat*
		Muamalat
4	Al-Mudharabah*	Mudarabah
5	Al-Murabahah*	Murabahah
6	Al-Musharakah*	Musharakah
7	as-sunnah*	Sunnah

*In Arabic the word 'al' and 'as' are literally translated as 'the'.

Printed in the United States
By Bookmasters